THE

ARYAN RACE

ITS ORIGIN AND ITS ACHIEVEMENTS

BY

CHARLES MORRIS
AUTHOR OF "A MANUAL OF CLASSICAL LITERATURE"

CHICAGO
S. C. GRIGGS AND COMPANY
1888

Reprinted 2004 by
Liberty Bell Publications
PO Box 890
York, SC 29745
www.libertybellpublications.com
803.684.4408

ISBN: 1-59364-021-8
Library of Congress Control Number: 2004096161

Liberty Bell's
Politically Incorrect Classics
Volume 2

Printed in the United States of America

PREFACE.

IT is our purpose briefly to outline the history of the Aryan Race, — that great and noble family of mankind which has played so striking a part upon the stage of the world; to seek it in its primitive home, observe the unfoldment of its beliefs and institutions, follow it in its migrations, consider the features of its intellectual supremacy, and trace the steps by which it has gained its present high position among the races of mankind. The story of this people, despite the great interest which surrounds it, remains unwritten in any complete sense. There are many books, indeed, which deal with it fragmentarily, — some devoted to its languages, others to its mythology, folk-lore, village communities, or to some other single aspect of its many sided story; yet no general treatment of the subject has been essayed, and the inquirer who wishes to learn what is known of this interesting people must painfully delve through a score of volumes to gain the desired information.

Until within a recent period the actual existence of such a race was not clearly recognized. A century ago there was nothing to show that nearly all the nations of Europe and the most prominent of those of southern Asia were first-cousins, descended from a single ancestor, which, not very remotely in the past, inhabited a contracted locality in some region as yet unknown. Of late years much has been learned of the conditions and mode of life of this people in their original home, and of their migrations to the point where they enter the field of written history. From this point forward the part played by the Aryans in the history of mankind has been a highly important one, and there is no more interesting study than to follow this giant from the days of its childhood to those of its present imposing stature.

Our knowledge of the condition of the primitive Aryans is not due only to studies in philology. The subject has widened with the progress of research, and now embraces questions of ethnology, archæology, mythology, literature, social and political antiquities, and all the other branches of science which relate particularly to the development of mankind. Enough has been learned, through studies in these several directions, to make desirable a general treatment of the subject, and an effort to present as a whole the story of that mighty race whose history is as yet known to the world only in disconnected fragments. The present work, however, pretends to be no more than a preliminary handling of this extensive theme, a brief popular exposition which may serve to fill a gap in the realm of literature and to satisfy the curiosity of the reading world until some abler hand shall grasp the subject and deal with it in a more exhaustive manner.

Any attempt, indeed, to tell the story of the Aryan race, even in outline, during the recent age of mankind would be equivalent to an attempt to write the history of civilization, — which is far from our purpose. But in the comparison of the intellectual conditions and products of the several races of mankind, and in the consideration of the evolution of human institutions and lines of thought and action, we have a field of research which is by no means exhausted, and with which the general world of readers is very little conversant. Our work will therefore be found to be largely comparative in treatment, the characteristics and conditions of the other leading races of mankind being considered, and contrasted with those of the Aryan, with the purpose not only of clearly showing the general superiority of the latter, but also of pointing out the natural steps of evolution through which it emerged from original savagery and attained to its present intellectual supremacy and advanced stage of enlightenment.

As regards the sources of the information conveyed in the following pages, we shall but say that all the statements concerning questions of fact have been drawn from trustworthy authors, many of whom are quoted in the text, — though it has not been deemed necessary to crowd the pages with citations of authorities.

In respect to the theoretical views advanced, they are as a rule the author's own, and must stand or fall on their merits. Finally, it is hoped that the work may prove of interest and value to those who simply desire a general knowledge of the subject, and may in some measure serve as a guide to those more ardent students who prefer to continue the study by the consultation of original authorities.

CONTENTS.

THE ARYAN RACE.

I.

TYPES OF MANKIND.

SOMEWHERE, no man can say just where ; at some time, it is equally impossible to say when, — there dwelt in Europe or Asia a most remarkable tribe or family of mankind. Where or when this was we shall never clearly know. No history mentions their name or gives a hint of their existence; no legend or tradition has floated down to us from that vanished realm of life. Not a monument remains which we can distinguish as reared by the hands of this people; not even the grave of one of its members can be traced. Flourishing civilizations were even then in existence; Egypt and China were already the seats of busy life and active thought. Yet no prophet of these nations saw the cloud on the sky "of the size of a man's hand," — a cloud destined to grow until its mighty shadow should cover the whole face of the earth. As yet the fathers of the Aryan race dwelt in unconsidered barbarism, living their simple lives and thinking their simple thoughts, of no more apparent importance than hundreds of other primeval tribes, and doubtless undreaming of the grand part they were yet to play in the drama of human history.

Yet strangely enough this utterly prehistoric and ante-legendary race, this dead scion of a dead past, has been raised from its grave and displayed in its ancient shape before the eyes of man, until we know its history as satisfactorily as we know that of many peoples yet living upon the face of the earth. We may not know its time or place of existence, the battles it fought, the heroes it honored, the songs it sang. But we know the words it spoke, the gods it worshipped, the laws it made. We know the character of its industries and its possessions, its family and political relations, its religious ideas and the conditions of its intellectual development, its race-characteristics, and much of the details of its grand migrations after its growing numbers swelled beyond the boundaries of their ancestral home, and went forth to conquer and possess the earth.

How we have learned all this forms one of the most interesting chapters in modern science. The reality of our knowledge cannot be questioned. No history is half so trustworthy. Into all written history innumerable errors creep; but that unconscious history which survives in the languages and institutions of mankind is, so far as it goes, of indisputable authenticity. It is not, indeed, history in its ordinary sense. It yields us none of the superficial and individual details in the story of a people's life, the deeds of warriors and the tyrannies of rulers, the conquests, rebellions, and class-struggles, the names and systems of priests and law-givers, with which historians usually deal, and which they weave into a web of inextricably-mingled truth and falsehood. It is the rock-bed of history with which we are here concerned, the solid foundation on which its superficial edifice is built. We know nothing of the deeds of this antique race. We are ignorant of the numbers of its people, the location and extent of its territory, the period of its early development. But we know much of its basal history, — that history which has wrought itself deeply into the language, customs, beliefs, and institutions of its modern descendants, and which crops out everywhere through the soil of modern European civilization, as the granite foundations of the earth's strata break through the superficial layers, and reveal the conditions of the remote past.

Such a germinal history of a people may very possibly lack interest. It has in it nothing of the dramatic, nothing on which the imagination can seize; none of those personal details or stirring incidents which so strongly arrest the attention of readers; nothing to arouse the feelings or awaken the passions and emotions of mankind. It has none of the ever-alluring interest of individual human life, — the hopes and fears, the joys and sorrows, the sayings and doings of men, great and small, which give to the gossipy details of history an attractiveness only a degree below that of the imaginative novel. Over our work we can cast none of this glamour of individualism. We have to do with man in the mass, and to treat history as a philosophy instead of as a romance. We are limited to the description of what he has done, not how he did it, and to the detail of results instead of processes. And yet history in its modern era is rapidly entering this philosophic stage. For many centuries it has been confined to the romance of individual life. It is now verging toward the philosophy of existence, the scientific study of human development. Kings and courtiers have too long dwarfed the people. But the stature of the people is increasing, and that of rulers and heroes diminishing, while a growing interest in the story of humanity as a whole is succeeding that in the lives of individuals. This gives us some warrant for venturing to describe the history of a race whose ancient life we know only as a whole, and of which we cannot give the name of one of its heroes, the scene of one of its exploits, or even the region of the earth which it occupied. Yet this race is so important a one, and its later history has been so grand and exciting, that the story of what is known of its primitive life can scarcely fail to find an interested audience, particularly when we remember that we are here dealing with our own ancestors, and tracing the pedigree of our own customs and institutions.

In this inquiry it is necessary to begin by considering the claim of the Aryans to the title of "race." What position do they hold in the category of human races, and what were the steps of their derivation and development from primitive man? We must locate them first as members of the broad family of mankind before we can fairly enter into the study of their record as a separate group. We have spoken of them somewhat indefinitely as a race, family, or tribe. Indeed, they cannot justly be honored with the title of race until we know more fully in what the race-characteristic consists, and what is their claim to its possession. In this respect ethnologists have so many varying ideas that the number and limitations of the human races are still far from being settled. We can therefore but briefly detail some of the latest views upon the subject.

Race-divisions, indeed, have been made through two widely different lines of research. Of these, the first and most fundamental is that of physical characteristics; the second is that of linguistic conditions. The latter, based on the radical diversities in human languages, doubtless indicates a more recent separation of mankind. To a considerable extent it follows the lines of physical variation. It seldom crosses these lines to any important extent, though it separates some of the broad physical divisions into minor races. The Aryan is one of these linguistic races. It is not a true race in the wider sense, since, as at present constituted, it includes portions of two physical groups which have so intimately intermingled that pure specimens of either are somewhat exceptional, and are found in any considerable number only on the opposite border-lands of these groups.

The primary separation of mankind into races very long preceded the development of the modern families of language, and was due to strictly physical influences. The mental lines of division, as indicated by language, are much more recent. The physical races have been variously classified by ethnologists, one of the latest schemes being that of Professor Huxley, who distinguishes four principal types of man, — the Mongoloid, the Negroid, the Australioid, and the Xanthochroic; to which he adds a fifth variety, the Melanochroic.[1] It is only with the last two of these that we are here directly concerned, since it is these which enter into the composition of the Aryan race. More recently Professor Flower has given an outline of a system of human classification which he regards as most in accordance with the present state of our knowledge on the subject.[2] He considers that there are three extreme types, — those called by Blumenbach the

[1] Journal of the Ethnological Society, ii. 404 (1870).

[2] Address before the Anthropological Institute, Jan. 27, 1885.

Ethiopian, the Mongolian, and the Caucasian, around which all existing individuals of the human species can be ranged, but between which every possible intermediate form can be found. Of these the Ethiopian is secondarily divided into the African Negroes, the Hottentots and Bushmen, the Oceanic Negroes or Melanasians, and the Negritos as represented by the inhabitants of the Andaman and other Pacific islands. The Australians, whom Huxley takes as the type of a separate race, he considers to be a mixed people, as they combine the Negro type of face and skeleton, with hair of a different type. His second race is the Mongolian, represented in an exaggerated form by the Eskimo, in its typical condition by most of the natives of northern and eastern Asia, and in a modified type by the Malays. Excluding the Eskimo, the Americans form one group, whose closest affinity is with the Mongolian, yet which has so many special features that it might be viewed as a fourth primary division. His third or Caucasian race includes two sub-races, — the Xanthochroic and Melanochroic of Huxley. The seat of this race is Europe, northern Africa, and southwestern Asia, its linguistic division being into Aryans, Semites, and Hamites.

Several recent writers are inclined to accept a conclusion closely similar to that of Professor Flower, and to divide man into three typical races, — the Negro, the Mongolian, and the Caucasian or Mediterranean; viewing all remaining races as secondary derivatives of these: as, for instance, the American and the Malay from the Mongolian; or as mixtures, as the Australians from the combination of the Oceanic Mongolians and Negroes. Topinard[1] goes so

[1] Anthropology, p. 510.

far as to divide man into three distinct species. The first of these is the Mongolian, distinguished by a brachycephalic, or short skull, by low stature, yellowish skin, broad, flat countenance, oblique eyes, contracted eyelids, beardless face, hair scanty, coarse, and round in section. The second is the Caucasian, with moderately dolichocephalic, or long skull, tall stature, fair, narrow face, projecting on the median line, hair and beard abundant, light-colored, soft, and somewhat elliptical in section. His third species is the Negro, with skull strongly dolichocephalic, complexion black, hair flat and rolled into spirals, face very prognathous, and with several peculiarities of bodily structure not necessary to name here.

It is not our purpose to express any opinion upon this theory of specific differences in mankind, except to say that if such differences exist they are probably limited to the Negro and the Mongolian stocks. There are good reasons for removing the Caucasian from this category. That the Negroes and the Mongolians do differ in sufficient particulars of structure to constitute a specific difference in the lower animals, must be admitted.[1] Their mental

[1] Agassiz notes the following marked differences in physical structure between the Negroes and the Indians of Brazil, — the latter in all probability originally of Mongolian race. His conclusions are based on the comparison of a large number of photographs of the two races. The Negroes are generally slender, with long legs and arms, and a comparatively short body; while the Indians have short arms and legs, and long bodies, which are rather heavy, and square in build. He compares the former to the slender, active Gibbons; the latter to the slow, inactive, stout Orangs. Another striking distinction is the short neck and great width of shoulder in the Indian, as compared with the narrow chest and shoulder of the Negro. This difference exists in females as well as males. The legs of the Indian are remarkably straight; those of the Negro are habitually flexed, both at hip and knee. In the Indian the

differences are equally marked. But these variations may possibly have had another origin. The Negro is essentially the man of the South, the developed scion of the African or the Australasian tropics. The Mongolian is the man of the North, his native region being the chill tablelands of northern Asia, so far as the balance of indications goes. Whether these two races, with their specific differences, arose as distinct species in these widely separated localities, and spread outward from these centres of dispersion until they met and intimately mingled at their borders, or whether they indicate some very early division of a single human species into two sections, and variation under differing climatic influences, are questions which science is not as yet prepared to answer. It is unquestionable that their well marked and strongly persistent physical characteristics are the outcome of a very long period of separate development. If there was a single primitive type of man, its two main divisions must have been long exposed to very diverse conditions of climate and life-habits; and its separation must have taken place at a very early era in human existence, — perhaps, as suggested by Professor Wallace,[1] at that primitive epoch when men were as yet too low in mind to combat against the influences of nature, and were far more plastic to the agency of natural selection than they have been during the later epoch of weapons, clothing, and habitation.

If we now come to the consideration of the Caucasian

shoulder-blades are short, and separated by a wide interval; in the Negro they are long, with little space between them. There are other differences of structure, equally marked; but the above will suffice to show the strong racial distinction. *Vide* "A Journey in Brazil," pp. 529–32.

[1] Contributions to the Theory of Natural Selection, p. 319.

race, we have to deal with a series of facts markedly distinct from those relating to the other two races named. In the Caucasian we certainly have not a primitive and homogeneous type of mankind, but a race of varied mixture and of much more recent origin, and therefore necessarily not a distinct species of man, but a derivative from primitive man.

In support of this view an argument of some cogency can be offered. The opening of the historical era presents the three races above indicated in very different relations to those which now obtain. At the earliest date to which we can trace them, the Mongolian and the Negro, with their sub-types and hybrid races, divided the major part of the earth between them. Hardly a foothold was left for the Caucasian. Great part of Africa and many of the Pacific islands were occupied by the Negro race. Others of these islands, all of America. and nearly all of Asia, were occupied by peoples of the Mongoloid type. As for Europe, late research has given us some very interesting information concerning its early inhabitants. There is reason to believe that it has been successively occupied by sections of the three principal human races, and that its general occupancy by Caucasians reaches not very remotely beyond the historical era.

The skull is the truest index of human races, and the ancient skulls found by modern man in Europe tell us much concerning its early ethnological conditions. The most ancient of these skulls belong to a long-headed, strongly prognathous race, with characteristics of a lower type than are to be found in existing man. This, called by Quatrefages the Canstadt race, includes the famous Neanderthal skull, with its brute-like characters. Other skulls, of apparently later date, constitute the so-called Cro-Magnon race. These are also dolichocephalic and prognathous, and approach nearer to the Negro than to any other of the existing types. It is not impossible that a modified branch of the Negro race had spread itself over western Europe at this early period.

Still later appear the skulls of men of quite different race-characteristics. These range from medium to short heads, while the accompanying skeletons are of short stature, and present certain traces of affinity to the modern Lapps. It is probable that the long-headed and possibly Negroid earlier race had been driven back by a Mongoloid migration, which in the Neolithic age became widely distributed. There are apparently two types, of which the medium-skulled one may be to some extent a cross between the long-headed aborigines and the intruding short-headed race. This "Neolithic" type has probably left a remnant of its language in the Basque dialect, as spoken by half a million of persons crowded into the Biscayan region of France and Spain, the relics of a people who once may have occupied the greater part of Europe. Though the language of Neolithic man has nearly vanished, his race-characters still persist; for the skulls and bodies of the ancient tombs seem reproduced in the physical characters of many of the present inhabitants of the same regions. The ancient race has held its own persistently against the later infusion of Aryan blood.

Thus in the outgrowth of what we incline to view as the two original races, the Mongoloid and the Negroid, the former seems to have been far the more energetic. It not only occupied the continents of Asia, Europe, and America, but pushed its way into northern Africa and the islands of the Pacific, yielding in the line of demarcation of the primitive races a type of man of intermediate characteristics. Though Mongolian man is less prolific than the Negro, his greater restlessness and spirit of enterprise seem to have placed him in possession, at a remote period, of most of the earth outside of Africa and the Asiatic islands.

In this glance at prehistoric man no clearly defined trace appears of the Caucasian race, whose area at that era was certainly very contracted as compared with that of the Mongolian and the Negro. And yet at the earliest date to which we can trace them the Caucasians exhibited the qualities they still possess, — those of superior intellectuality, enterprise, and migratory vigor. When we first gaze upon the race,— or rather upon its Xanthochroic section,— it is everywhere spreading and swelling, forcing its way to the East and the West with resistless energy. Before its energetic outflow the aborigines vanish or are absorbed. In the continent of Europe no trace of them is left, with the exception of the Basques, pushed back into a mountain corner of Spain, and the Finns and Lapps, driven into the arctic regions of the North. A similar fate has befallen them in southern Asia. During the whole historical era this migratory spirit has continued active. The separate branches of, and the Aryans as a whole, have been persistently seeking to extend their borders. They are still doing so with all the old energy, driving the wedge of invasion deep into the domain of Mongoloid and Negroid life, until the Caucasians of to-day number one third of all mankind,[1] and bid fair, ere many centuries, to

[1] About 420,000,000. Two centuries ago their number was not more than one tenth of the earth's population.

reduce the other races to mere fragments, like the Basques or the North American Indians of the present day.

From these facts we certainly have some warrant to conclude that the Caucasian is not a primitive human race, but a peculiar and highly endowed derivative of the preceding races. Otherwise we should not have found it at the beginning of authentic history almost lost in the sea of ruder life, but its superior qualities would have told at a far more remote epoch, the Negro and the Mongolian expansion have been checked long ages ago, and history opened with the Caucasian as the dominant race of mankind. It is generally acknowledged that from the primitive types many sub-races have branched off, differing in mental and physical characters; as, for instance, the American from the Mongolian. The Caucasian may possibly be a very divergent example of these sub-types, or rather, if we may judge from certain highly significant indications, a compound of two sub-types derived from the two preceding races.

Of the two sub-races which make up the Caucasian stock of mankind, the Xanthochroi, or fair whites, are now found most typically displayed in the north of Europe, mainly in Denmark, Scandinavia, and Iceland. The Melanochroi, or dark whites, have their typical region in northern Africa and southwestern Asia. Between these regions an intimate mixture of the two types exists, endless intermediate grades being found; though as a rule the Xanthochroic becomes more declared as we go north, and the Melanochroic as we go south.

The combined race is described by Peschel[1] in the following terms: The shape of the Caucasian skull is

[1] The Races of Man, p. 481.

intermediate between the short skull of the Mongolian and the long skull of the Negro race. Prominence of the cheek-bones and prognathism, or projection of the lower jaw, common characters in the other races, are very rare in the Caucasian, or the Mediterranean race, as he names it. The skin varies in hue. Fair hair and blue eyes with a florid complexion are very frequent among the Northern Europeans. Such was also the case with the Gallic Celts, as described in ancient history, though it is not so with the modern French, with whom the darker hue prevails. The skin is generally darker with the Southern Europeans, and becomes yellow, reddish, or brown in Africa and Arabia, while the hair and eyes become dark or black. The hair of the Mediterraneans is not so long nor so cylindrical in section as in the Mongolians; it is not so short nor so elliptical as in the Negroes. It is generally curly, being intermediate between the other two races in this respect. The hair is more abundant than in the other races, and the beard much more so, the Mongolians and Americans being nearly beardless. The nose is a well-marked feature, its high bridge and narrow form distinguishing it from the broad and flat nose of the Negroes and Mongolians. The lips are usually thin, and never present the swollen aspect of the Negro lips. As a whole, the features of this race are more refined than those of the other races, and the form is more symmetrically developed.

The Caucasian, indeed, seems as a rule intermediate between the other two races. The Negro face, seen in profile, recedes from the chin to the forehead; that of the Caucasian is vertical. The Mongolian face is vertical or projecting in profile, but in front view is of a triangular outline, being broad at base and contracted at the forehead; the Caucasian outline is oval. The flat median line of the Negro and the Mongolian is replaced by a projecting outline in the Caucasian, mainly due to the elevation and narrowness of the nose and the lack of expansion in the cheek-bones.

In these particulars the two sub-races of the Caucasian somewhat closely agree, their main distinction being in color, though there is also a marked difference in form. The Xanthochroic, or blond type, is distinguished by blue or gray eyes, hair from straw-color to chestnut, and a rosy or florid complexion, which burns to a brick-red or becomes freckled under exposure. In form this race is tall and stout, of square build though sometimes slim, with rather ponderous limbs, and a squarer skull and coarser features than in the Melanochroic.

The latter race is marked by a skin of brownish or olive hue, which quickly blackens upon exposure, sometimes enormously so; it perhaps inherits a tendency to revert to the typical Negro complexion. The color of the hair and eyes is black, and the stature lower than in the Xanthochroi. The form is very symmetrical in its proportions, the skull round-domed, and the features are more delicate than those of the blond type. These two types, as we have said, have become intimately mingled, so that every shade of gradation exists between them. Yet numerous instances of the typical structure appear, and the race-characteristics seem very persistent.

The blond race has its purest expression in Iceland, Scandinavia, and Denmark, and next in Holland, northern Germany, Saxony, Belgium, and the British Islands. But it crops out throughout the whole range of the Caucasian domain. In the far East, though the brown type is generally prevalent, the blond type frequently appears. It is common among the Persians and Afghans, while the Siah Posh of Kaffiristan are particularly marked by their fair complexions, blue eyes, and chestnut hair. It exists also in northern Africa, and on an Egyptian monument of the twelfth dynasty there appears the representation of a man with white skin, blond hair, and blue eyes. Yet in this southern region the dark type is the prevalent one, while it in its turn has forced its way far to the north, though in diminishing frequency as it approaches the colder regions.

The natural inference from these facts is that the blond type has its native locality in the North and East, in contiguity with the Mongolian, and the dark type in the South, in contiguity with the Negro race. The expanding tendency which these types of man have displayed during the whole historical epoch must have existed since their first origin, if we may judge from their very intimate commingling, which has been so great that comparatively few pure representatives of either type remain. No such complete mixture is shown in the Mongolian and Negro races, except in a narrow border region. This indicates a much less energetic constitutional migratory spirit in the latter than in the Caucasian, and is a further argument in proof of the recent origin of this race; since if of remote origin, it could not possibly have been confined to the narrow region in which we find it at the opening of the historic period.

What, then, was the origin of the two Caucasian sub-races? In response to this question we may propound the views offered by Mr. J. W. Jackson,[1] who advances the theory

[1] Aryan and Semite, Anthropological Review, vii. 333.

that the Semite (or, as we prefer to consider, all the Melanochroi) is really a derivative from the Negro race; and the Aryan (or rather the Xanthochroi) is a derivative from the Mongolian. He bases this theory on mental characteristics; but he should have considered also the physical characters of the races. If we observe the Melanochroi, or dark whites, it is to find their purest specimens in the far South, on the immediate northern limits of the Negro race. And here they present significant points of affinity to the Negro type. Many of the Berbers of the Sahara region approximate to the Negro in feature, though some tribes are light olive in complexion, with straight noses and thin lips. Of the ancient Egyptian type we are told that they had "thick lips, full and prominent; mouths large, but cheerful and smiling; complexions dark, ruddy, and coppery; and the whole aspect displaying — as one of the most graphic delineators among modern travellers has observed — the genuine African character, of which the Negro is the exaggerated and extreme representation."[1] The Arabs present similar affinities. Some of the Arab tribes of the Middle Desert have crisp hair, approaching that of the Negroes in texture. In bodily and mental character the Southern Arabs of pure blood approximate to the Negro type,[2] and in color they may become of a jet black, as is the case with the Shegya Arabs of Africa. On the other hand, in northern and more elevated regions the complexion of the Arabs is as fair as that of Europeans.[3] Quatrefages looks upon this

[1] Denon, Voyage en Égypte.

[2] Palgrave, article "Arabia," Encyclopædia Britannica (ninth edition).

[3] Prichard, Natural History of Man, p. 150.

race as one which has evolved a single step beyond the "arrested" Negro phase.[1]

Tribes of mankind closely affiliated with the Melanochroi, though with a stronger infusion of the Negro element, extend much farther south in Africa. In addition to the Melanochroic Abyssinians and Gallas, may be mentioned the more Negroid Nubas, with black skins, but features of a type intermediate between the white and the black races. But the most significant of the mid-African peoples are the Foulahs, — an energetic and warlike tribe, distinctively different from the Negroes, into whose domains they are steadily intruding. This people has become much modified by intercrossing with Negroes and Arabs, but seems to have been originally of the Melanochroic type. Dr. Lenz, in his recent work on Timbuktu, says of them that they are of a distinctly non-Negro type. Pure specimens of the Foulahs differ from the Negroes in almost every racial characteristic, — in cranial conformation, complexion, texture of hair, figure, proportion of limbs, and in mental qualities. He was amazed at their striking resemblance to Europeans, and describes the pure-blooded Foulahs as of light complexion, slightly arched nose, straight forehead, fiery glance, long black hair, shapely limbs, tall, slim figures, and of great intelligence.

In fact, the Melanochroi present indications, to judge from their early wide extension, of being a much more primitive race than the Xanthochroi. They are found throughout northern Africa, extending to a line drawn considerably south of the Sahara; widely distributed throughout southern Asia, from the Semitic regions to India, where they give the main physical character to the Hindu Aryans;

[1] The Human Species, p. 351.

everywhere in southern Europe, where their type greatly predominates over that of the blonds; and in less preponderance in central Europe, where they have essentially modified the original type of the Celtic and Teutonic Aryans.

If we accept the indications here presented, in connection with the apparently very limited extension of the blond type of man in the recent pre-historic period, we are led to the theory that the Eastern Hemisphere was divided at a more remote period between three races of mankind, — the Mongolian in the temperate and frigid zones, the Negro in the tropics, and the Melanochroi occupying a broad intermediate belt stretching across the whole continent from the Atlantic to the borders of Farther India.

It is interesting to perceive that this zone occupied by Melanochroic man is that of demarcation of the primitive Mongoloid and Negroid races. Here they must have met and mingled, and here a hybrid derivative of the two races very probably arose, — an intermediate type of mankind, with a preponderance of the Negro element, if we may judge from existing indications. It is particularly in Europe that we find evidence of this mingling of the long-headed and short-headed aboriginal races, their resultant being a type with skulls of medium length, — the Neolithic man of western Europe. More extended investigation may yield similar evidence all along the zone of demarcation. We can picture to ourselves an original Negroid population in this zone, a southward migratory movement of the more enterprising Mongolians, and a long-continued mingling of the two races, with a somewhat profound modification of their physical characteristics, yielding a new type of man, the Melanochroic, with considerably more of Negro than of

Mongolian blood, yet essentially diverse in character from both the parental types.

If now we come to consider the origin of the blond type of man, we find ourselves brought down to nearly historic times. The widespread extension of this type at the opening of the historic era can be traced back, almost step by step, to an original central region, probably of small dimensions, though of unknown location. We have evidence from the Egyptian monuments of what may have been the first appearance of blond man in that region. Of the type as found in the north of Africa, in Tunis and Morocco, among the Berbers of the Sahara, and in the Canary Islands, Topinard remarks: "It is derived from a Tamahou people who about the year 1500 before our era made their appearance upon the frontier of Egypt, coming from the North. . . . The blonds which we meet with in the Basque territory and near the Straits of Gibraltar in Spain are probably descendants of theirs."[1] In Europe and Asia the movements of the blond race took place immediately before the opening of the historic epoch; and though the centre of dispersion is not clearly known, yet nearly every step of migration has been traced. In every region to which they migrated, with the exception of Scandinavia, they seem to have mingled freely with the preceding Melanochroic inhabitants, yielding that intimately mixed race which constitutes the Aryan of to-day. To this fusion we owe the modern man of southern Asia and Europe, from the bronzed Brahman of the East to the round-headed and dark-featured class among the Celts of the West. Only in the extreme North did the Xanthochroic type sustain itself in any purity, and only in Arabia and Africa did the

[1] Anthropology, p. 452.

Melanochroic type remain preponderant. In all the region between, every possible intermediate gradation of the two types exists, though the dark type gradually decreases as we move northward, and the blond type as we move southward.

If we endeavor to seek the derivation of the blond type of man the indications are very obscure. This type differs markedly from the Mongolian; and yet we are not without intermediate links of connection, or traces of a tendency in the Mongolian to assume the Xanthochroic characters. We are told by Chinese historians of certain mysterious tribes in central Asia who were tall of stature and had green eyes and red hair. Matuanlin, the historian, described one such people as inhabiting western Mongolia at the opening of the Christian era. A similar tribe existed beyond the Altaï Mountains. Other tribes are mentioned, down to the twelfth century, as tall, with red hair and green eyes, and of fair complexion.

Some writers are inclined to consider these as members of the Turkish Mongolians, who are known to have inhabited the region mentioned. The physical appearance of the modern Turks, indeed, strongly resembles the Aryan type of man. The Turks of the Ottoman and Persian empires are completely Europeanized in feature and structure. This is by some ascribed to persistent intermarriage with Circassian slaves; yet such a theory applies only to the rich and powerful, while the peasantry are equally Europeanized. The great mass of the lower population have always strictly intermarried, difference of religion and manners keeping them separate from the Greeks and Persians. The Tadjiks of Persia, the true Aryans, are of a sect of Mohammedanism hostile to that professed by the

Turks, and these two classes have kept rigidly separate. The Aryan characteristics of the civilized Turks is therefore not so readily explainable.

Of the Turcomans Vambéry says that they alone of all Mongolians do not possess high cheek-bones, while the blond color is predominant among them. Yet the Turkish hordes of the northern steppes are strongly Mongolian in physical character, though occasionally blue and gray eyes are observed among the Kirghiz. Still farther eastward similar indications appear. Topinard quotes as follows: "We saw Mantschu Tartars," says Barrow, "who accompanied Macartney's embassy to Pekin, men as well as women, who were extremely fair and of florid complexion; some of the men had light blue eyes, a straight, aquiline nose, brown hair, and a large and bushy beard."[1] All this, however, might be due to mixture with the blond race, even though we have no evidence of conditions favorable to such a mixture. Yet such could not well be the case in America, where similar variations are common. King tells us that "the oval face associated with the Roman nose" is by no means rare among the Eskimos, while the complexion is sometimes fair, sometimes dark. Among the American tribes the nose is occasionally of the Mongolian type, but is often large, prominent, bridged, and even aquiline, while the stature is tall, and the skull has a tendency to the elongated shape. Several tribes, both of North and South America, present a close approximation to the European type. This is strikingly the case with the Mandans, the so-called White Indians of the West, as described by Catlin. The above facts seem to indicate a ready variability in the Mongolian race, under the influence

[1] Anthropology, p. 452.

of diversity of climate and condition, since these widespread modifications towards the European type can scarcely be ascribed to mixture with a race as limited in numbers as the Xanthochroi appear to have been at the opening of the historic era.

There is yet, however, one branch of the linguistic Mongolians to be considered, — the Finnish. And here we find a strongly marked approximation towards the Xanthochroic race, far too general to be ascribed to intermarriage. The Finns are to some degree intermediate between the blond and the Mongolian types, though much nearer the former. They are marked by long hair, usually reddish or yellowish, or of a flaxen hue, and more rarely chestnut. The European Finlanders have red hair, with a moderately full beard, generally red. The eyebrows are thick, the eyes sunken, and of a blue, greenish gray, or chestnut hue. The complexion is fair, and usually freckled. The nose is straight, with small nostrils; the cheek-bones are prominent, owing to the thinness of the face; the lips small. These characteristics clearly separate the Finns from all the surrounding types, and bring them much closer to the European than to the Mongolian race. The northern Russians in particular are of very similar physical character. Very probably the green-eyed and red-haired race spoken of by the Chinese were Finnish tribes, though blue is more common than green in the eyes of modern Finns. We may also say here that the Finns approach the Aryans in the possession of a mythology and of a highly developed poetry, — an evidence of mental power which is not found in pure Mongolians of a similar state of civilization.

Thus though no direct clew to the origin of the Xanthochroic type of man exists, there are strong indications that it was a derivative from the Mongolian, and that it arose at a comparatively recent date. We have shown that a tendency exists among the Mongolians of northern Asia and America to deviate towards the Xanthochroic character. In the case of the Finns this deviation has yielded a strongly marked race, nearly approaching the Xanthochroi both physically and mentally. It is of interest, in this connection, to remark that the Finnish race is native to a locality bordering upon that which the latest archæologists consider the original home of the Aryans, and that it differs from the neighboring Russians mainly in language, and very little in physical character. It may be offered as a conjectural hypothesis that the primitive Xanthochroi were a derivative from the Finns at an era before the languages of either had attained much development, the further physical variation which took place being probably due to climatic influences, and possibly to residence of the Xanthochroi in a mountainous region.[1]

The mental characteristics of the several human races lead us to similar conclusions. In the first place it may be remarked that all the savage tribes of the earth belong to the Negro or the Mongolian race. No Negro civilization has ever appeared. No Mongolian one has ever greatly developed. On the other hand, the Caucasian is pre-emi-

[1] It seems probable that the Lapps, the remaining European Mongolians, have close race-affinities with the Finns. Professor A. H. Keene has recently examined a company of seven Lapps, in London, and decides that in several respects they have deviated from their fundamental Mongolian type, and have assimilated, especially in the color of the hair and eyes, in the complexion, and in the shape of the nose, to the surrounding Norse population. He attributes this assimilation to like climatic influences rather than to intermixture, of which there is no direct evidence. The family belonged to the mountain nomadic tribes, of purest descent and of least intercourse with Europeans.

nently the man of civilization. No traveller or historian records a savage tribe of Caucasian stock. This race everywhere enters history in a state of advanced barbarism or of rapidly advancing civilization.

But the Caucasian development is not the work of either of the sub-races, but of their combined resultant. Mentally, each of the pure types too closely approaches its assumed ancestral race to display vigorous intellectual powers. The pure Melanochroi tend towards the Negro type of intellectuality; the pure Xanthochroi approximate to the Mongolian. The Negro race, as described by De Gobineau,[1] is marked by a low grade of intellectuality, combined with a strongly emotional tendency. It is quick in acquisition at first, but soon stops, and grows dull intellectually. Emotionally the Negro is capable of violent passions and strong attachments. He has a childish instability of humor, intense but not enduring feelings, poignant but transitory grief. He is seldom vindictive, his anger being violent but quickly appeased, his sensibilities ardent but speedily subsiding. His amatory feelings are strong, and his sensuality highly developed. In these particulars he is akin to the Melanochroi of Arabia and the West, in whom we find a sensual temperament, fierce passions, intense emotions, and a mentality that requires excitement more than reason for its exercise, and tends to the fanciful far more strongly than to the logical.

If now we compare the yellow race with the black, we find them strongly opposite in mental characteristics. In muscular vigor and intensity of feelings the typical Mongolians are greatly inferior to the blacks. They are supple and agile, but not strong. Their sensuality is less violent

[1] Moral and Intellectual Diversity of Races, p. 445.

than that of the blacks, but less quickly appeased. They are much less impulsive, and rather obstinate than violent in will-power. Their anger is vindictive, but not clamorous. They are seldom prone to extremes, and while easily understanding what is not very profound and sublime, their lack of emotional and imaginative energy prevents their attaining an ardent faith or an exalted religious philosophy. They love quiet and order, and keenly appreciate the useful and practical. They are, indeed, a practical people in the narrowest sense of the word. Their lack of imagination renders them uninventive, but they easily understand and adopt whatever is of practical utility.[1] This description applies mainly to the Asiatic Mongolians, and is shown in the whole conditions of the Chinese civilization. It cannot be extended to include the Americans, who have a very marked development of the faculty of imagination. It applies in some measure, however, to the blond race of northern Europe, in whom we find a strong mental antithesis to the ardent nations of the South. The pure blonds replace the nervous temperament of the Melanochroi with a lymphatic temperament. They lack vivacity, but are more reflective. They are controlled by reason rather than by desire. Conclusions are not reached impulsively, but are thought out, and are strongly held when once arrived at. They are not of quick passion, are slowly roused, but earnest and persevering, and are brave without requiring the stimulus of enthusiasm. They are sincere and simple-minded, but addicted to gluttony and drunkenness, — faults to which the Melanochroi are much less addicted. In these respects the blond white presents the same affinity to the Mongolians as the dark white does

[1] Moral and Intellectual Diversity of Races, A. de Gobineau, p. 445.

to the Negroes, and they seem respectively the highest expression of these two races.

But in the mentality of the two primary races we have the germinal conditions of the highest phases of intellectual development. The emotional characteristics of the Negro are the germinal stage of the imaginative faculty; the practical mentality of the Mongolian is the germinal condition of the reasoning powers. In Scandinavia we find a practical people, yet one not given to abstract thought. In Arabia and northern Africa we find a highly emotional people, yet one not noted for valuable imaginative productions. For the higher unfoldment of these mental faculties a further step was needed, — that close fusion of the two sub-races which has so widely taken place. The mixed race of Europe presents us with the highest type of man. The wild flights of Southern fancy have been tamed by the cool decisions of practical sense, until we find, as the lineal successor of the Oriental extravagance, the artistically imaginative productions of the people of Greece. The practical tendency of the Northern mind has been inspired by imagination until it has yielded the exalted products of Teutonic reason.

Despite the long and close intermingling of these sub-races, the mental character of each crops out frequently in strong isolation, now reason, now imagination, becoming markedly predominant in an individual or a people. The highest display of the reasoning faculty in modern Europe is in the region of the Teutonic race, in which the infusion of Xanthochroic blood is in excess. The imaginative faculty has reached its highest development in the South, where Melanochroic blood is in excess. This is markedly displayed in the literature of Greece, and yet more so in

India, where the flights of imagination have left reason far in the rear. In mid-Europe of to-day these two faculties exist in some degree of balance: though in France and the South the preponderance of imagination is shown in the artistic and picturesque tendency of thought, while in Germany a like preponderance of the logical faculty appears; and in England, the central meeting-place of the two races, these two faculties seem more evenly combined than elsewhere upon the earth. It is to this mingling of South and North, of fair and dark, of judgment and emotion, of imagination and reason, that we owe the Aryan race, the apex of human development, and the culminating point in the long-continued evolution of man.

The comparative mental characteristics of the three typical human races are briefly enumerated by De Gobineau in the following terms: The white race has great physical vigor, capacity, and endurance. It has an intensity of will and desire which is controlled by intellectuality. Great things are undertaken readily, but not blindly. It manifests a strong utilitarianism, united with a powerful imagination, which elevates, ennobles, and idealizes its practical ideas. The Negro can only imitate, the Chinese only utilize, the work of the white; but the latter is abundantly capable of producing new works. He has as keen a sense of order as the yellow man, not from a love of repose, however, but from the desire to protect and preserve his acquisitions. He has a love of liberty far more intense than exists in the black and yellow races, and clings to life more earnestly. His high sense of honor is a faculty unknown to the other races, and springs from an exalted sentiment of which they show no indications. His sensations are less intense than in either black or yellow, but his mentality is far more developed and energetic.

Our hypothetical line of human physical development may be combined with one of mental development in a brief synopsis of the progress of human mentality. Very far back in time it is possible that a single race of man occupied the earth, brute-like both in body and mind, if we may judge from the most ancient traces of mankind yet discovered. At a later epoch two strongly marked races made their appearance, perhaps as derivatives from the single primeval race. Or, in the opinion of some, these two races were primitive, and constituted two original species of man. They differed essentially both physically and mentally. The Negro race was marked by a strong emotional tendency, in consonance with its tropical climate; the Mongolian by an equally strong phlegmatic and practical mentality, in consonance with its frigid climate. At a much later date these races gave rise to two more highly developed types of man, — the Melanochroi, in which the Negro emotion had unfolded into imagination, and the Xanthochroi, in which the Mongolian practicality had developed into logic. Finally, an intimate mixture of these two sub-races yielded the modern dominant type of man, the Aryan, in whom logic and imagination have become combined into reason and art, and the special, one-sided mental development of earlier man has become a generalized, intermediate condition of mentality which can be most fairly characterized by the title of intellectuality. Thus the Aryan stands as the type of intellectual man, the central outcome of the races, in which the special conditions of dark and light, North and South, emotional and practical, have mingled and combined into the highest and noblest states of mind and body.

If now we come to consider the lines of race as indicated by language, they will be found to follow to some extent those above given, though they separate mankind into several minor racial divisions. The considerable diversity in physical character between the Americans and the Asiatics, for instance, indicating, as it does, an early separation, is in conformity with the indications of language, since each continent has its strongly marked linguistic type. Linguistically the Caucasians are divided into three sub-types, — the Aryans, the Semites, and the Hamites. Between the first two of these the distinction in language is very decided. Between the Semites and the Hamites it is much less declared, and their types of language seem to have grown up in close contiguity. Significantly, these latter types of language are spoken by peoples of Melanochroic blood. But no Xanthochroic people has ever been found speaking any but an Aryan tongue.

II.

THE HOME OF THE ARYANS.

IN seeking to trace the original home of the Aryans we are concerned mainly with the Xanthochroic, or blond, type of the race. The Melanochroic, or dark, type was widely spread, in the later prehistoric era, throughout the Mediterranean and the southern Asiatic region. But the blonds were in all probability far more limited in locality, and their place of residence remains one of the unsolved problems of science, despite the persistent efforts which have been made to discover it. Yet these blonds or "fair whites" were the true Aryans, the people with whom the type of language known as Aryan originated. The languages of the "dark whites" belong to a very distinct family of speech, which is still spoken by most of the typical representatives of the race, though Aryan tongues are generally spoken by the tribes and peoples arising from a mingling of the two races. It is therefore the original home of the Xanthochroi — the blue-eyed and fair-haired ancestors of the modern Aryans — that we shall here endeavor to trace.

The effort to solve this problem has mainly been based upon considerations of comparative philology. It has been a fascinating pursuit to its devotees. The speech of the original Aryans was wholly unknown; yet fragments of it lay buried in the depths of modern language, and these have been assiduously wrought out and pieced together, until, like an edifice built of disjointed materials, they yield a complete and coherent image to our minds. Word by word the language of the ancient Aryans has been exhumed. But a word represents a thing, a relation, or an action, and points to some possession or activity of the people who used it; and the words of a language embody the whole industrial, social, and political life of a nation, down to its minutest detail. Unfortunately we do not know the language of the ancient Aryans in any such complete sense as this, nor are we quite sure what meanings they attached to their words. Yet their study has given us some very interesting glimpses into the lives of a vanished people, and enabled us, to some extent, to bring them back again to the surface of the earth.

The discovery that a close affinity exists among the languages of Europe is a result of very recent research. The resemblance between Greek and Latin, indeed, has long been known, and the common descent of the Romanic languages, — the French, Spanish, and Italian, — was too evident to be lost sight of. But that the remaining languages of Europe were first-cousins of these, was not perceptible until philology had become a science. The divergences, though of the same character, were much wider than those between the Romanic languages, and needed a critical study before the resemblance could be made apparent.

Ere this work had made any important progress another and very distant language was brought into the same family. The English in India had become acquainted with the Sanscrit, — the noble and venerable language of the Vedic literature of the Hindus. To their surprise and delight, they discovered that this interesting language possessed close links of affinity, both in words and in structure, with the European family of speech. This was first pointed out by Sir William Jones about 1790, who declared that the three languages, the Latin, Greek, and Sanscrit, had sprung from "some common source, which perhaps no longer exists." He was also inclined to attribute the Persian to a similar source, and hinted at the possibility of the Celtic and the Gothic being members of the same group.

This earliest conception of an Indo-European family of languages was taken up and extended some twenty years afterwards by Frederick Schlegel, who in 1808 maintained the theory that the languages of India, Persia, Greece, Italy, and Germany were connected by common descent from an extinct language, just as the modern Romanic tongues were descended from the Latin. For this vanished dialect he proposed the name Indo-Germanic. The truth of this theory was first demonstrated by Bopp, in his "Comparative Grammar," published from 1833 to 1852. He not only proved clearly the close affinity in grammatical structure between the languages above named, but also added the Zend, Armenian, Slavonic, and Lithuanian to the group. The Celtic dialects were included about the same time; and the relationship of all the members of the great family of Aryan speech was thus made evident. For this group the name "Indo-European" was proposed, — a name which is still used by many philologists. The term "Aryan" has more recently come into favor, mainly through the influence of Max Müller. This title really applies only to the Persians and the Hindus, being that by which they knew themselves before their separation; yet its shortness and ease of handling is giving it ascendency over the complex compound titles as a name for the whole widely extended family. Systematic philologists have entered into long arguments to prove that the word "Aryan" has no right to be applied to all Indo-European peoples. No one disputes the validity of these arguments, and yet the proscribed word has come generally into use. It is short and convenient; and this is of tenfold more importance to ordinary speakers than its etymology. To make a close research into the origin of words is one of the tasks of philology; but this does not carry with it the necessity of replacing accepted and convenient terms by more correct but cumbrous synonyms. In all languages there are thousands of words whose origin is quite lost in their application; philologists are aware of their original signification, and nothing further is required.

The community of origin of the peoples above named had been suspected from other lines of study long before this linguistic demonstration was completed. Ethnologists and mythologists had lent aid to the demonstration. A connection between their religious ideas had become evident, and the similarity of their race-characteristics had been observed. Dr. Pritchard suggested their affinity, from a study of their skulls, years before it was proved from a study of their languages. But the results of these earlier investigations were only partially accepted, and the work of the philologists was needed to round out the circle of proof. This evidence from philology was no light task. The separation of the Aryans into distinct branches had taken place so long ago, and the language of each branch had so diverged from those of the others, that it was not easy clearly to prove their relationship. But science is patient and persistent; it has long sight and clear vision. One by one the difficulties vanished, and the truth was made apparent. One of the most striking forms of linguistic divergence was that pointed out by Jacob Grimm and met by the celebrated "Grimm's law." He showed clearly that each branch of the Aryan family had peculiar tendencies of speech, resulting in certain variations of vowels and consonants, which were constant for the same people. Whether from some change in the vocal organs that rendered one letter more easily pronounced than another, or from some unknown cause, each nation developed its own peculiar variations from the original Aryan sounds, so that a single primitive word often assumed forms quite unlike in sound, and seemingly incompatible in form. Thus the consonant sound that became *v* in one branch of the Aryans became *b* in another. *S* with this people became *th* with that. Here the vowel was aspirated, and there the initial *h* was suppressed. Several such methods of change might be named, each dialect branching off in its own special direction, the German following one line, the Latin another, etc. It is the discovery of the system of vocal change prevailing with each people that constitutes Grimm's law, and that enables us to prove the identity of words which at first sight seem to have nothing in common. As one illustration of this we may quote Max Müller's identification of the English word *Nelly* with the *Saramá* of the Vedas. The *s* in Sanscrit often becomes *h* in Greek, and the liquid *r* as often becomes *l*. Thus Sanscrit *Saramá* became Greek *Halama*. This, by an ordinary Greek modification, became contracted to *Halan*. But the Sanscrit *a* is often changed to *e* in Greek, and by such a change *Halan* became *Helen*. The further steps of change were easy. *Helen* in English has become *Ellen* by the loss of the aspirate, and *Ellen* has become transformed into

Nelly as a familiar name. Yet between these two words of the same origin there is not a single letter in common. Philologists do not often have to handle such intricate tasks as this; yet their labors have been by no means trifling, and the above will serve as an extreme instance of the changes with which they have had to deal.[1]

It will suffice here to say that this line of inquiry has been carried to the point of absolute demonstration. There is no more doubt entertained to-day by scientists of the original community of the languages of the peoples named than there is of the existence of the earth. The proof does not rest upon a possibly chance resemblance of words, but deals with the very nerves and sinews of speech, — that rigidly persistent grammatical structure which survives the most radical changes in the forms of words. These separate peoples, as Whitney remarks, all count with the same numerals, call individuals by the same pronouns, address parents and relatives by the same titles, decline nouns by the same system, compare adjectives alike, conjugate verbs alike, and form derivatives in the same method. The words in most ordinary use are similar in them all. The terms for God, house, father, mother, son, and daughter, for dog, cow, heart, tears, and tree, are of the kind that would naturally persist. No chance could produce abundant conformities of this close charac-

[1] We may give, as an illustration of the verbal community of the Aryan languages, the forms taken by one or two words in the several tongues. Thus the word "house" is in Sanscrit, *dama* or *dam;* in Zend, *demana;* in Greek, *domos;* in Latin, *domus;* in Irish, *dahm;* in Slavonic, *domu:* English derivative, *domestic.* In like manner, "boat" in Sanscrit is *nau* or *nauka;* in Persian, *naw* or *nawah;* in Greek, *naus;* in Latin, *navis;* in old Irish, *noi* or *nai;* in old German, *nawa* or *nawi;* in Polish, *nawa:* English derivative, *nautical.*

ter between a whole series of languages; and the general existence of such conformities absolutely demonstrates the common origin of the Aryan tongues.

But a demonstration of the common origin of languages leads to that of the common origin of the peoples who speak them. If there was one original Aryan language, there was one original Aryan people. It does not follow, however, that the modern speakers of Aryan tongues are all descendants of this people. Oppert, Hovelacque, and other able philologists claim that the correspondence of Aryan languages does not prove a common descent, but is the result of the propagation of a language from a single centre through heterogeneous populations, as the Romans and Arabs spread Latin and Arabic over regions inhabited by other races. This theory, as originally advanced by M. Oppert, is vigorously contested by Professor Whitney. He cannot imagine that any circumstances existed in the early barbaric period similar to those of the Roman and Arabian empires. In his view, no aboriginal language has ever been entirely dispelled without a complete incorporation of the people; and this has never taken place except in the Roman empire. Nothing of the kind appears in the conquests of the Persians, Germans, Mongols, or even of the Greeks, and certainly could not arise in a much less developed people. The complete political and social fusion of the conquered with the conquering people of the Roman empire has never been paralleled in history, and existed only in those regions that were bound to Rome for many centuries. The Arabic parallel is a very imperfect one; it represents an infusion of the Arabic rather than an abolition of the native languages. Barbarians do not conquer in this complete way; they destroy or enslave, or their conquests end, after a limited period, in a revolt of the conquered tribe. Race-mingling may take place, but hardly an acceptance of the language of a conquering tribe by unamalgamated peoples. This argument of Professor Whitney is not, however, in very strict agreement with what race-indications tell us concerning the Aryan peoples. There can scarcely be a doubt that, in some instances, the vigor of the Aryans sufficed to impose their language on more numerous aboriginal peoples, with whom they became thoroughly mingled. Such, for instance, is the case with the Celts, the Slavonians, and the Hindus. There is much reason to believe that in all these the original Aryan conquerors mingled their blood with that of a considerably more numerous conquered people. Yet the Aryan language has held its own with very little modification, while the aboriginal speech has vanished. Certainly the vigor, enterprise, and persistent spirit of the Aryan migrants must have exerted a strong influence upon the more yielding aborigines, and we cannot be surprised if the latter often lost their language with their nationality.

We have sufficiently considered in the preceding section the question of the mingling of the "fair whites" and "dark whites" of Europe, and endeavored to show the probability that the development of this type of mankind, with its distinctive family of language, took place in a region distinct from that of the typical Melanochroic people. Where was this region? On what area of the earth's surface was it that the Aryan-speaking people grew into social, political, and linguistic coherence, and developed that budding civilization and migratory energy which were, at a later period, to send them forth to conquer the world? This is a question which has caused deep heart-burnings among philologists, which is yet far from settlement, and which may perhaps never be fully solved. Yet the early and hasty conclusions have been succeeded by better based and more consistent theories; and it is possible that the "home of the Aryans" may yet be determined with some satisfactory degree of approximation. The present state of this much-vexed question we shall briefly endeavor to set forth.

In the study of Aryan antiquity the languages of Europe present us only with words. No historical details or traditions exist to show an early migration from some remote locality. But in the eastern branch of the Aryan family there is abundant evidence of a migration to India and Persia. Literatures, reaching back beyond the date of this migration, exist, comprising the Vedic hymns of the Hindus, and the religious works of the Zoroastrian sect, in which some historical and geographical details are preserved. These indicate the region of ancient Arya, the common home of the Hindus and Persians while they yet formed a single people, or of all the Aryans, as was long maintained.

The theory of an eastern home of the Aryans was first advanced by J. G. Rhodes in 1820. Thirty years ago this home of the common Aryan tongue was supposed to be, in the words of Pictet,[1] the "vast plateau of Iran, that immense quadrilateral stretching from the Indus to the Tigris and Euphrates, from the Oxus and Jaxartes to the Persian Gulf." But this area was soon found to be too extensive, and attempts were made to reduce it within

[1] Les Origines Indo-Européennes, ou les Aryas Primitifs, p. 35.

more probable limits. The traditions of the Avesta seemed to point to the region of Bactria as the place of common residence of Hindus and Persians while they still formed one people. At that period, too, much was said about the plateau of Pamir, the "roof of the world," as the birthplace of the civilized races, though it is now clearly perceived that this inaccessible and inhospitable highland is utterly unsuited for human residence. In fact, the Avestan traditions were plainly stretched too far. They indeed contained reminiscences of an older Iranian land, but gave no warrant for the view that this land was the cradle of the whole Aryan race. Philology was next appealed to, and the claim made that the language which had most faithfully preserved the ancient Aryan type must have been the one that had migrated the least. This primitive condition was found in the Sanscrit and the Zend, while the Celtic, which had made its way farthest West, had apparently suffered the greatest transformation.

To the above conclusions, however, several objections may be made. In the first place, the fact that the early Persian and Hindu literatures indicate a migration, while no distinct tradition of the kind exists in the literatures of early Europe, proves, if it proves anything, that the eastern Aryans were the only migrating members of the race. And their comparatively small numbers and limited area in their early days is an evidence in the same direction. It is far more probable that the migration of a tribe from the West to the far East took place, than that the bulk of the race moved from the East to the far West, leaving a single tribe behind. And that these eastern Aryans were immigrants who forced themselves among hostile strangers, is abundantly indicated in their literature. It is a literature of battle, of deadly fray, of unyielding hostility. The Vedas are the stirring hymns of a people surrounded by strangers alien in race and religion, with whom there can be no peace, and whose destruction is a duty to God and man. They breathe the tone of an invading race full of vigor and bent on conquest. The Hindus seem to have been then, as they are to-day, plunged into the heart of an alien population. The Eastern Aryans have expanded much since those early days, but they are still everywhere surrounded by Mongolian tribes. India is still largely inhabited by members of the Mongolian race and by tribes of other race-affinity, while its pure Aryans are comparatively few. This relation obtains also to some degree in Persia and the other Asiatic Aryan districts. The vital Aryan stock has held its own, but it has had to contend with an alien multitude, and a great degree of mixture of races has necessarily taken place.

The argument from philology seems no more cogent. In the Vedas and the Avestas we have preserved to us relics of an early stage of Aryan speech which no longer exists as a living language in Asia, and has no counterpart in the languages of Europe. Had we remains of the latter from a period of equal antiquity, they might prove equally primitive. And that the Celtic has undergone the extreme transformation assumed, is questioned by recent philologists. In fact, the great probability is that the Aryans before their dispersion occupied a somewhat wide locality, into which they had gradually spread from their original contracted domain. As a consequence, their common speech must have undergone many changes and corruptions among the various tribes during the ante-migration period. Bopp found signs of many such derangements and disturbances in the organism of the original Aryan speech, seeming to show that they had dwelt in their early home for a long period after the primary development of their linguistic method. As they spread, dialectical changes necessarily increased, and quite likely the peculiar dialect of each branch of the race had become partly formed before the era of dispersion. Thus the argument from special primitiveness of any of the surviving modes of speech can scarcely be maintained. We know far too little of the diversities of speech in ancient Arya and of the early form of the languages of modern Europe to be able to come to any definite decision on this controverted point.

In fact the theory that the original Aryan home was in Bactria is no longer held except by the older philologists. The arguments upon which it was based have proved insufficient to sustain it, and no new ones have been advanced. Another line of argument, to which little attention was formerly paid, has led several recent writers to place in Europe the ancient Aryan home. It was suggested, early in the century, that the Slavonic was a primitive European population. More recently it has been claimed that Europe was the original seat of all the Aryans. This theory is maintained by H. Schulz, D'Halloy, Latham, Benfey, and others of the more recent writers, and is rapidly becoming the prevailing view. It trusts for its proof mainly to linguistic arguments.

Every word which is now used by all the Aryan peoples is considered to be a direct descendant from the antique speech of the race, and to indicate some ancient knowledge or possession of the Aryans. A study of these words gives us much interesting information as to the conditions of the original Aryan home. For instance, there is no common word for camel. The word in use has been borrowed from the Semitic languages. This seems decisive against Bactria, where the camel is an ordinary animal, and must have received a name of Aryan origin had the Aryan languages been formed in that region. In like manner no name for the lion or the tiger is common to the Aryan languages, and the inference is that the ancient Aryans were ignorant of these animals. To this it is objected that very many words must have been lost, and that these may have dropped out and been replaced by other terms. Yet such a conclusion is not based on probability. Many words far less likely to persist have been retained, and it cannot be reasonably maintained that the names of these terrible and destructive wild beasts would have been utterly forgotten, if once known. Yet if there were no lions or tigers in the primitive Aryan home we must seek this home in Europe, since these animals are found throughout southern Asia.

In this connection we may quote Peschel's views as to the original home of the Aryans, which are based on somewhat narrow grounds, it is true, yet have strong arguments in their favor in addition to those which he gives. "It lay eastward of Nestus, now Karasu, in Macedonia, which in the time of Xerxes was the limit of range of the European lion. It was still farther north than Chuzistan, Irak Arabi, and even than Assyria, where lions are still to be met with. It cannot have included the highlands of west Iran and the southern shores of the Caspian Sea, for tigers still wander in search of prey as far as these districts. Hence, from all the facts here cited, every geographer will agree that the Indo-Europeans occupied both slopes of the Caucasus, as well as the remarkable gorge of Dariel, and were in the habit of visiting either the Euxine or the Caspian Sea, or perhaps both. . . . It is usually objected to this argument that in the course of their migrations the Aryan families abandoned the territory of the lion and the tiger, and with the animals forgot their names also. But this requires stronger evidence, for the Maori have preserved the names for the domestic pig and the cocoanut, although neither existed in New Zealand. Had the ancient Aryans seen or fought against such magnificent animals in their own country, their names would certainly have been retained, even though with an altered significance."[1]

Other writers are inclined to place the Aryan home in the plains of southern Russia, and still others on the shores of the Baltic or in Scandinavia. In evidence of these hypotheses they present the following facts: The Aryans occupied a cold region. Of the seasons they have names only for winter, spring, and summer. Autumn was not recognized as a separate season. But the best series of common names for climatic phenomena are those belonging to winter. Cold and snow were well known. It was a freezing and shivering home in which our ancestors dwelt. Their dress consisted of tunic, coat, collar, and sandals. These were formed of wool or leather. Abundant provision was needed against the wintry chill. Among their wild animals were the bear and the wolf, among their common trees the birch, — all natives of the European temperate zone. They seem to have been unacquainted with the ass and the cat, — ancient domesticated animals of Africa. This indicates that they were too far removed from Egypt to have any intercourse with this very ancient civilization.

[1] The Races of Man, by Oscar Peschel, p. 507.

That they were acquainted with some large inland body of water, is admitted. They had boats, which they moved by oars. They had names for salt, and for crabs and mussels; but the oyster was unknown to their language, and they knew nothing of the ocean. The salt lake on which they made their maritime excursions is supposed by the Asiatic advocates to have been the Caspian. Those who advocate the Caucasian region, or the plains of southern Russia, suppose it to have been the Caspian or the Black Sea, or both. Those who place them in northern Europe point to the Baltic as their sea.[1]

Other evidences that Europe was the original Aryan home may be drawn from their historical distribution. At the earliest dawn of history they were found in possession of all Europe, except the frozen regions of Finland and Lapland in the extreme north. All Europe is named with their names, except where the geographical titles of the Basques persist. There is nothing to indicate that they are intruders, as in the case of the eastern Aryans. All tradition makes them natives of the regions where found. When first seen in history they are moving to the east and the south, not to the west.

As to the extreme migratory theory of Aryan dispersion, it can hardly be sustained. There is no evidence in its favor in the history of human migrations. The only tribes in the history of mankind which have completely released their hold of their early homes, and poured out *en masse* in search of a new home, have been pastoral peoples, with

[1] Late advocates of this theory are Professor Penka, who finds the ancient Aryan home in Scandinavia, and Professor Schrader, who locates them in northeastern Europe. Professor Sayce, noticing the works of these writers, considers the neighborhood of the Baltic the most probable region.

the possible exception of the legendary American migratory movements of hunting tribes. In Europe and Asia such complete migrations can be traced only to the pastoral tribes of Arabia and Mongolia; there is no record of any such movement of an agricultural people, such as the Aryans had become in considerable measure at the period of their supposed dispersion. That such a people could have flowed out in several great successive waves of complete migration to remote distances, is hardly credible, and is utterly without warrant in the history of human movements.

The Arabian outbreak of the Mohammedans was not a migration in the complete sense. It was a swelling beyond the national borders, incited by hope of plunder and desire for religious propagandism. Arabia continued the centre of the movement, and the only settlement made in a region remote and disjoined from this central home was that formed in Spain. This instance presents a suggestive parallel to that of the eastern Aryan branch, with its pious horror of the impious tenets of its foes, and its wide separation from its kindred race.

Yet the primitive Aryans, while advanced in great part beyond that nomadic pastoral stage of industrial life which has been the condition of all migrating peoples known to history, had not yet reached that degree of political consolidation and religious culture requisite for definite invading movements *en masse* for the purpose of propagandism. It seems far more probable, therefore, that the movements of the Aryans were expansions rather than migrations, — the incessant bite of restless and enterprising tribes into the domains of surrounding peoples. As their numbers increased, and their primitive home became too small to hold them, they may have pushed out in this manner in all directions with the restless energy which has always characterized them, driving back the original populations before their resistless expansion. This idea would seem to indicate an original home in some such central region as that suggested by Peschel, midway between the eastern and western extremities of the Aryan outflow, and offering easy roads for expansion alike to the East and the West.

The majority of the recent authors, however, seem inclined to accept the Baltic or the Scandinavian region as the primeval Aryan home. Of the several arguments offered in support of the latter hypothesis the most potent one is the fact that Scandinavia is the only region of the earth now occupied by pure Xanthochroi, who lose their typical characters more and more as we advance southward, until they are quite lost in the strong preponderance of Melanochroic blood. But this is by no means a convincing argument. The degree of mingling with the aboriginal inhabitants depended very much on the numbers of these inhabitants and on the character of their treatment by their conquerors. Either strong resistance or strong race prejudice might have resulted in their annihilation or their complete dispossession. The only Scandinavian aborigines of whom we have any knowledge are the Lapps, — a Mongolian people with whom the Aryans have shown no inclination to mingle, and who may originally have been driven back to the frozen plains which they at present inhabit. The Xanthochroic purity of the Scandinavians can be accounted for quite as well on this as on the other theory. The Germans and the Celts of Gaul were of equally pure Xanthochroic blood as recently as the times of Cæsar and Tacitus. Their loss of purity of type is due to a mixture since that period with the Melanochroic aboriginal element. No such mixture appears to have taken place between the Scandinavians and the Lapps.

A potent argument against the Scandinavian theory is that the Aryans were a pastoral people in the early era of the formation of their language, and partly pastoral at the period of their migrations, their domesticated animals, with the exception of the camel, being the same as those possessed by the nomads of the Asiatic steppes. No pastoral people has ever originated except on broad, open levels, with abundant pasturage, — a condition which the Scandinavian peninsula does not present. Hunting and fishing habits were the only ones likely to originate in that wooded and seagirt land, except in the far North, where the snowy levels gave an opportunity for the use of the reindeer as a domesticated animal. But this native Scandinavian beast of burden does not seem to have been known to the primitive Aryans, — which would certainly not have been the case had it been used by them or their immediate neighbors. As the lack of a common word for the camel has been used as an argument against Asia, so the similar lack of a common word for the reindeer tells against Scandinavia as the primitive home of the Aryans.

Nor does the region of the Baltic or the levels of northern Russia answer any better to the requirements of the case. It is not simply a land which the Aryans might have inhabited in accordance with the indications of philology, but one that is in harmony with their mode of life and process of development, that we seek; and this can certainly not be found in a densely wooded region, such as the Baltic provinces were in primeval times.

At the period in which the Aryan method of speech

began to deviate from the Mongolian (to which it has the closest affinities of type), and Aryan man to deviate perhaps from the Finnish division of the Mongolian race (which most closely approaches him in structure), the habits of the Aryans appear to have been purely pastoral, and probably long continued so. This is clearly indicated by the character of the root-words of their languages. The balance of probabilities, therefore, favors their residence in a locality of Europe contiguous to that occupied by the pastoral Mongolians and the Finns, and one naturally well adapted to pastoral pursuits.

A brief study of the development of mankind shows us that the pastoral habit has originated nowhere except on the broad open plains and deserts of Asia and of northeastern Africa. No such pursuit has ever been followed in mountain districts or forest regions. And the animals possessed by the nomadic Aryans were those indigenous to Asia, with the exception of the camel, which is suited only to sandy deserts. If the home of the pastoral Aryans was in Europe, it must have been in a locality adapted to this mode of life and contiguous to the Asiatic steppes. The only European region which properly fulfils these requirements is that of southern Russia. The remainder of Russia and of northern Europe was then, and is yet in considerable measure, a dense forest; while southern Europe westward of this region is, from its mountainous character, absolutely unfitted for the life of the nomad shepherd and herdsman. But the region of southern Russia, particularly in the vicinity of the Caspian, is an open level plain, partly desert, partly of high fertility, and presenting the requisites of contiguity to the Asiatic steppes, the primeval home of the wandering herdsman, and of excellent adaptation to pastoral pursuits. It is simply impossible that such pursuits could have originated or been maintained in a forest country, nor is it conceivable that the barbarians of that age had the means or the inclination to clear the land of forests for the purpose of providing pasturage.

The next subject of consideration is the fact that the Aryans gradually lost their nomadic habits, assumed a settled state of existence, and began to practise agriculture, which in time they developed to an extent that rendered their pastoral pursuits of secondary importance. Their locality must have been one suited to this change of industrial habits. An inquiry into the requisites for the development of agriculture is therefore here in place.

Again we must leave the forest and seek open and naturally fertile regions. So far as we know or have satisfactory reason to believe, agriculture in the Eastern Hemisphere originated only in localities specially favored by nature. It arose on the highly fertile banks of the Nile, of the Tigris and the Euphrates, of the Ganges and the Indus, and on the rich lowlands of the great rivers of China. There were agricultural districts elsewhere in Asia, it is true; but it is probable that these localities derived their knowledge of the art from the regions named, and not from a spontaneous development. In America similar indications present themselves. The agriculture of the United States region not improbably arose on the rich border-lands of the lower Mississippi, and was disseminated northward by the Mound-Builders. Like conditions probably attended its origin in Mexico and Peru.

There is, in fact, not a particle of evidence in existence

that agricultural habits ever originated spontaneously in a cold forest region such as that of the Baltic, while this region was too far removed from the agricultural districts of Africa and Asia for the art to be gained through commerce or instruction. Such a region, while utterly unadapted to pastoral pursuits, is equally unsuited to the gradual exchange of these for agricultural conditions. In short, the only pursuits which appear to have ever naturally arisen in forest-covered countries are those of the hunter; with those of the fisher where large bodies of water are contiguous. And as respects the districts of northern Germany, what we know of the habits of the tribes in the days of the Roman empire indicates that they were not only disinclined to agricultural progress, but that they showed a tendency to neglect the agricultural knowledge they already possessed, and to revert to the hunting stage, so well suited to thèir forest surroundings.

On the contrary, the region of southern Russia and the Caucasus, from its openness, its fertility of soil and suitability of climate, and its contiguity to the Syrian district of Asia, from which the art of the agriculturist might have been readily gained, seems particularly well adapted to the gradual change from pastoral to agricultural pursuits, particularly within the limits of the mountain range, which the expanding nomads would naturally have penetrated, and which were unsuited to the life of the herdsman.

There is still one matter of importance to consider. We have given what seem to us satisfactory reasons for the belief that the Xanthochroi are not an original race of mankind, but a derivative from a preceding race, in all probability from the Mongolian, and that their origin dates from a somewhat recent period. Yet the development of a new type of feature and new structural conditions of body could hardly have taken place in regions similar in physical character to those native to the parent race. We have seen that this race frequently assumes a type of face and complexion closely approaching the Aryan; but such a tendency could not well have a general development except as due to a marked change in physical surroundings and conditions of life, as in the case of the American Indians and the Mongolians of northern Europe. In the instance of the Aryans the change may have been due to residence in a mountainous district such as that of the Caucasus. In such a region, with its great difference in climate, physical surroundings, and necessary life-habits and industries from life on a plain, a marked change in structure might well have taken place, while the conditions of existence might have necessitated a gradual development of that art of agriculture which was already practised in the neighboring district of southwestern Asia.

For the various reasons here given, and others which will be advanced in the next chapter, we incline to look upon southeastern Russia as the home of the Aryans during their nomadic era, and the Caucasian mountain region as the locality in which they gained their fair complexion and the other characteristics of the Xanthochroic type, perfected the Aryan method of language, learned the art of agriculture, and developed their political and religious ideas and organization.

From this mountain stronghold, in which they could well have sustained themselves against all aggression during the long period of their development as a distinct people, they probably spread into the fertile plains of southeast Russia, occupying the district between the Caspian and the Sea of Azov, and extending an indefinite distance northward and westward. Their northern borderlands may have been the home of the primitive Russians, since these deviate less from the Mongolians than any other section of the Aryans, and bear to-day a close resemblance in physical aspect to the Finns. Had the Aryan type of language been imposed upon the Finns, and the latter thus been classed as an outlying member of the race, we should have an almost unbroken line of deviation, leading from the typical Xanthochroi to the Mongolian type of man.

The region we have indicated as the primitive home of the Aryans has a further point in its favor. This is its propinquity to the Semitic populations of the South, and the ease with which the fair and dark types might have mingled in that early stage of culture which preceded strong political and religious antipathies. It seems a natural point of meeting of the highest outcome of the races of the North and the South, and may have much to do with the existing strongly Melanochroic character of the southern Aryans. And to it may be due that strong invigoration of the Aryan intellect, by the infusion of the imaginative element of the Southern mind into the practical groundwork of Mongolian mentality, which was necessary to the unfoldment of its high powers of thought and to the development of the energy which has carried the race with unflagging persistence outward from its narrow primeval home to the conquest of the world.

At a later period came the development of property rights, of the exclusive Aryan system of clanship, and of religious bigotry and fanaticism; and with it a strong feeling of hostility to strangers, and a rigid effort at isolation, such as we find in similar historical cases. Such conditions would have checked the infiltration of alien blood, and given an opportunity for the full development of the Aryan type of speech and of social, political, and religious institutions undisturbed by foreign influence.

Scarcely a trace of such influences appears in the language and institutions of the Aryans; and whatever its steps of origin, the Aryan, in all the details of structure and in mental character, is among the most distinct and declared of human races, and is markedly separated from all other tribes and divisions of mankind.

III.

THE ARYAN OUTFLOW.

IF we look back through time to the most remote point to which the scope of history or tradition extends, it is to behold Europe and Asia the scene of active movement and endless turmoil. Everywhere tribes, communities, nations, are in motion, extending their borders, overrunning one another's domains, battling for the choice spots of the earth, thirsting for the wealth which the industry of the more civilized holds out to the avarice of the more barbarous. It is everywhere the same. Alike in Italy and Greece, in Syria and Babylonia, in Persia and India, in China and Scythia, the tribes and nations are moving with the bewildering confusion of a phantasmagoria. It is to us a shifting of names rather than of peoples. Numerous titles of tribes have descended to our times, but we know very little of the communities which these names represent; and the surface of the earth at this early epoch appears to us like that of a chess-board on which meaningless figures are incessantly moving to and fro. Of only one thing we can be sure. We are aware of the general race-relations of these migrating peoples. We know that the movements in Europe and in southern-central Asia are mainly Aryan, while the Syrian movements are Semitic, and those of northern Asia are Mongolian. Of the migratory excursions of the period in question much the most extensive are the Aryan, the movements being wider, and the hold upon new regions more decided, than in the case of the other races of mankind.

But that this condition of affairs is representative of the whole scope of human history, from the earliest date of man's appearance upon the earth until the present time, can hardly be affirmed. Such a migratory spirit has existed throughout the period of recorded history, but its results have been steadily growing more extensive during the progress of civilization. The movements which our earliest records present to us are minor in character. We perceive migrations of small tribes to short distances, in place of the subsequent marches of great armies over thousands of miles. Such is the character of the early migratory movements and hostile excursions as recorded in the Bible, and of the similar movements of the Italian and Grecian tribes. Such was also the case with the military enterprise of the primitive civilizations. The records of the early dynasties of Egypt and Babylonia yield no evidence of extensive operations. The story of ancient China is that of the battling of tribes. Nor was this growing empire as yet exposed to any serious danger from the pastoral hordes of the North, who had not yet learned the art of moving in mass.

The limited enterprise which we thus behold at the opening of history, as compared with the extensive movements of a later period, is significant of a still more diminished migratory activity in the prehistoric ages. The spirit of outflow had perhaps just become active, and the mingling of the races but fairly commenced, when historical records begin. In fact a considerable degree of intellectual advancement is necessary to any active enterprise of this character. We find nothing of the kind among the savage peoples of the earth. The savages of to-day make no effort to extend their domains. Each tribe naturally spreads until it reaches the borders of another tribe, and there it rests in dull contentment. This border-line is usually a line of hostility, but not of energetic movements of invasion. In Africa, for instance, we hear of no migrations of the full-blooded Negro tribes. Activity is confined to the Foulahs and other mixed races. That much movement took place in the early epoch we have good reason to believe, from the evidences of a very ancient occupation of the whole earth. But this was perhaps largely due to human fecundity, not to human enterprise. From the original centre or centres of population man slowly spread out, as his numbers increased, to occupy the earth, with only the difficulties of nature and the hostility of wild beasts to check his outflow. This expansion may have taken many thousands of years for its completion. But when the earth was once fully occupied, a strong check took place. Everywhere man met man. Doubtless an incessant hostility ruled, but nothing existed which we can properly term aggressive war. Each tribe or race remained confined to its ancient domain, with but slow and unimportant widening or shifting of borders. Only those peoples who by a greater advance in intellect had become superior in arms and in enterprise, slowly spread outward, gradually pushing back their weaker and duller neighbors.

The views here offered are in accordance with the facts indicated by the existing condition of human races. We are aware how great a mixture of races has taken place since the opening of the historic period. Pure races are in the minimum, mixed races are in the maximum, throughout the earth. And this is particularly the case in the regions of greatest civilization. It is strongly displayed in southern Asia, and still more strongly in southern Europe. For any near approach to purity of race in a people we must seek the regions of barbarism and savagery, mainly the locality bordering on the Arctic Circle, and the tropics of Africa and America. Had an energetic migratory and invasive spirit existed during the long centuries of the human past bearing any close relation to that of the early historic period, a complete mixture of mankind must have taken place, and the existence of well-marked races to-day would have been impossible. Race-distinctions would have been obliterated, as they now are to a great extent in the centres of active civilization. The epoch of the rise of an active migratory spirit, then, is one of great importance in the history of mankind. This epoch was probably the one immediately preceding the birth of recorded history, if we may judge from indications. We see evidences of such a spirit in the early history of China, Babylonia, and Egypt, probably considerably preceding its appearance among the Aryans. And yet the latter, when once they entered the circle of migratory activity, speedily became the most enterprising of human races. There are reasons for these conclusions in the history and conditions of these several races.

The industrial and political condition of the Aryans greatly differed from that of the Semites and the Mongolians. The latter were nomadic pastoral peoples. The Aryans, though strongly pastoral at first, became to some extent agricultural at a remote date. The indications are that they were not nomadic in the period immediately preceding history, and that they were divided into a great number of small groups. This we judge from their political system, that of the Village Community, which must have been long in developing, and which indicates a protracted period of fixed residence and agricultural habits. As a result of this system they were greatly inferior in political consolidation to the nomad tribes of the desert. Each of these formed a single group. The Aryans were divided into many small groups, diverse in their interests. The desert tribes were accustomed to rapid and extensive movements, in which they carried their property with them. The Aryans were tied to their property, which consisted, in part, at least, of fixed soil, and not entirely of moving herds, as with the nomads. And, finally, the organization of the nomad tribe was that of an army. It was under its single sheik, or patriarchal leader, who directed all its movements, and who might at any time set in train an invading enterprise. The Aryan organization was that of a community of equals. It was thoroughly democratic, and only by a slow process of development did it come under the control of warlike chiefs or leaders. It was not invasive, though it probably held its own vigorously against invasion.

From this difference in condition we can understand the difference in the history of the agricultural and the nomad peoples. The nomads of the northern and southern deserts, while perhaps inferior, even then, to the Aryans in intellectual vigor and in industrial development, were far better adapted for migratory movements and for the invasion of neighboring regions. This doubtless explains the invading movements in China, Babylonia, and probably Egypt, and the establishment of powerful agricultural kingdoms in these localities under a form of government closely analogous to that of the pastoral hordes of the desert, while yet the Aryans remained in a barbaric state, slowly advancing industrially, but almost stagnant politically.

The subsequent difference in the historical development of these races is due to the fact that the Aryan political organization is one that admits of steady unfoldment, while that of the pastoral races is essentially primitive and unprogressive. The only change the latter are capable of is the extension of the rule of an able chief from a single tribe to a wide circle of tribes, — to which we owe the terrible Mongolian migrations of the Middle Ages. Yet these could produce no important permanent effect, since they lacked any strong principle of political consolidation. The Aryan principle, on the contrary, was one which but slowly developed, with the increase of authority in the tribal chief, but it was one that depended much less on able leaders than on vitality of organization. Thus the Aryan movements have been persistent instead of occasional, and their effects permanent instead of transitory. Where the Aryan sets his foot, there he stays. There have been some temporary yieldings before the wild onslaught of feebly combined pastoral hordes; but these have in nearly every instance been recovered from, and the Aryan movement has been and is steadily onward, driving back before its firm front all the other races of mankind.

If now we come to consider particularly the outflow of the Aryan race from its primitive home, we must begin by seeking to trace its condition and relation to other tribes at that epoch. As to the locality of this home, we have given what seems to us the most probable of the several theories; namely, that it was in the region of southeastern Europe, stretching from the Black to the Caspian Sea, and probably northward to a considerable distance over the level steppes of Russia, with their chill climate and their excellent natural adaptation to both pastoral and agricultural habits. Southward it may have occupied the range of the Caucasus, and perhaps have crossed this range and extended some distance into the mountainous district to the south.

In addition to the reasons already given for this hypothesis, it may be remarked that it would be difficult to select a region better adapted to be the cradle-spot of the future conquerors of the earth. No district in Europe or Asia is better protected against invasion. With broad seas to the right and the left, and a lofty mountain-chain to the south, passable only at two easily-defended points, it is only approachable from the north. In the early days of the race, when it may have been stationed in close contiguity to and within these mountain-fastnesses, it could have defied all invaders, as the modern Caucasian mountaineers so long defied the power of Russia. Here developing in stature, in physical conformation, in intellect, and in habits of settled life, of agricultural industry, and of democratic organization; and here perhaps receiving a new spirit of enthusiasm through partial amalgamation with the Melanochroic peoples of the South,— the typical Aryan race originated, as we conceive, and began its outflow in a slow movement northward over the flat and fertile plains which stretch away from the very foot of the Caucasian chain.[1]

[1] It may be said here that a movement of this precise character has prevailed throughout the historic period among the Russian agricultu-

At a date preceding that of the more active migratory movement, this slow preliminary growth northward may have spread the Aryans over a district of considerable extent, and already divided them into several distinct and mutually hostile branches, with dialectical variations of language and marked peculiarities of custom. The system of language doubtless originated while the race was contracted in locality and numbers. The dialectical variations arose after its expansion. The skeleton of Aryan speech was the same in all the subsequent branches, yet considerable superficial differences existed. Possibly the Celtic, the Teutonic, the Greco-Italic, the Iranic, and the other main stems of Aryan speech had already strongly declared themselves while yet the race remained a compact body, its outermost branch still in the vicinity of the primeval home.

At this period the region which the Aryans were afterward to occupy was in the hands of alien races. Southern Asia, from Armenia to India, was held by tribes partly Mongolian, and partly perhaps of Melanochroic race. So far as India is concerned, we know this to have been the case, from the very abundant remains of the aborigines yet existing. In Persia, Afghanistan, etc., there are fewer traces of the aborigines; they have mainly perished or been incorporated with the conquerors. In Europe the only existing distinct communities of the aborigines are the Lapps and Finns of the North, and the Basques of the Southwest. All the remaining aborigines have sunk

rists, and still persists. There is plentiful room for expansion in that broad land, and the farmers seek new localities as necessity or fancy dictates. This migratory spirit has been made use of by the Russian Government to colonize their newly conquered lands.

beneath the Aryan tide, though it seems certain that much amalgamation has taken place. In fact, at the very beginning of European annals the domain of the Aryans seemed nearly as extensive as now. We have no clear trace of the aboriginal inhabitants. Several names survive, such as Pelasgians, Leleges, Amazons, Iberians, and Aborigines, as the titles of ancient Mediterranean populations; but just what these names indicate, no one can positively declare. The Pelasgians were possibly an early Aryan tribe of migrants, though this lacks satisfactory evidence. The Iberians are now taken as the clearest representatives of the ancient European race. The Etruscans of Italy may also have been members of this race; but the remnants of their language are too scanty to admit of a decision, and it is held by many that they were Aryans.

Of the nearly mythical peoples named, the title of Iberians was applied by the old geographers to the pre-Aryan inhabitants of the peninsula of Spain and the southwest of France, whose final remnant is supposed to exist in the Basques. But everything in relation to the Iberians is exceedingly uncertain. We now know, however, that an aboriginal people, the Neolithic, or users of polished stone implements, of small stature, with round or oval skulls, occupied this region at a remote period, and extended into Britain, Belgium, Germany, and Denmark. They resembled the Basques physically more than any other living people of that region, and possibly extended into Africa and formed part of the Berber population. This was probably the antique European element, semi-savage or barbarous in condition, with which the Aryans came into contact, and which they partly annihilated and partly absorbed. Indications of such an amalgamation exist in the historic Celtiberians of Spain, — a supposed mingling of the Celts with the Iberians. Other indications exist in the small, dark type of man found to-day in Aquitania and Brittany, and also in Wales, in the Scottish Highlands, and in parts of Ireland.

As to the localities occupied by the branches of the Aryan people in the period just preceding the era of invasion, some tentative suggestions may be made. As above said, the race probably occupied a considerable district, and comprised several distinct and perhaps hostile divisions. Of these, that which we now know as the Celtic was the most westerly in situation, the most divergent in language, and possibly the most hostile in feeling towards its kindred. The Teutonic branch probably occupied the most northwesterly situation, the Indo-Iranian the most southeasterly, and the Greco-Italic the most southwesterly, while the Slavonic occupied the central and northern regions. This conjecture is mainly based on what we know of the directions and dates of march of the different branches, and partly upon another circumstance. This is that the northerly portion of the population would naturally be least exposed to the influx of Melanochroic blood, and the southerly portion the most so. Thus the typical Xanthochroi would be specially found in the border regions to the north and west, — those here ascribed to the Celtic and Teutonic branches. It is in the Teutonic branch that the typical Xanthochroi are still mainly found, and particularly in its frontier portion, — that which made its way to Scandinavia. As for the adjoining Slavonians, their most northerly section, the Lithuanian, is to-day distinguished by the fair hair and blue eyes of the Xanthochroi from the darker Russians of the South. On the other hand, the Indo-Persian branch is strongly Melanochroic. This is also the case with the Greco-Italians. As for the Celts, they are known to have presented originally a strong display of Xanthochroic characters, though these have been lost through their subsequent amalgamations.

There is, therefore, reason to believe that all the northern Aryans — the Celts, Teutons, and Slavonians — were originally of the pure blond type, and very little affected in their native home by admixture with an alien element. This may be deduced from the fact that all the early historians describe them, after the date of their migration, as a large-framed, blue-eyed, fair-haired people. The strong probability is that their present diversity of type resulted from intermarriage with Melanochroic and Mongolian aborigines at a comparatively recent period. In the geographical scheme we have adopted, this section of the primitive Aryans occupied the fertile plains extending northward and westward from the Caucasian range. The southern section, the Greco-Italic and the Indo-Iranian, which may have occupied the southern portion of the range and the mountainous district farther south, would be in a position to mingle freely with the Melanochroi of Armenia, Asia Minor, etc., before their migration. Their present strongly declared Melanochroic character may be due mainly to such an antique intermixture, and in a lesser degree to subsequent admixture with the aborigines of their later homes.

It is not improbable that the Celts led the vanguard in the great Aryan march. In fact they had begun to meet the fate of their dispossessed foes at the opening of the historic period, and were, being more and more crowded into the most westerly portions of the European continent by later invaders of their own race. The incitement to their first movement we shall never know. Probably the Aryan giant was growing beyond the dimensions of its natal home, and needed more space for its developing limbs. More than one of the historic migrations has been due to a pressure from behind, as in the case of the Huns. Such a hostile pressure may have set the Celts in motion, and, indeed, may have kept them in motion, it proving easier to overcome the uncultured aborigines in front than to endure the Aryan pressure from the rear. The movement of the Celts seems to have been always one of onward push, if we may judge from what is known of their history.

The Celtic was probably the easiest of the Aryan migrations. It met with less capable foes, as we may conjecture, than the eastern migration, while all subsequent European invasions had Aryans to deal with, and therefore found a far more difficult path to victory. When this first outflow took place it is impossible to guess. It may, and may not, have been far back in the prehistoric era; and it is impossible to say how many centuries were occupied in the movement. The Aryans were yet learning the art of invasion. They had not the arms or the military skill of the later migrants. Their progress was possibly a very slow one. As for the extant history of this Celtic migration, it may be outlined in a few words. When first we become acquainted with the Celts, they occupied a very extensive district, comprising most of Europe west of the Rhine, and the domain of Cisalpine Gaul in northern Italy. They had probably long before crossed the Channel and settled the British Islands.

But Spain appears still to have been held by the aborigines.

The earliest of the Celtic military movements of which history tells us was that famous one, under the lead of Brennus, which captured the young city of Rome, and but for a chance in the chapter of accidents might have stifled that scorpion in its birth. A century later another Brennus led a Gaulish force far to the east, which ravaged Thrace, pillaged the Grecian temple of Delphi, and received from Nicomedus, king of Bithynia, a settlement in Asia Minor, in the district called after them Galatia. After having met the ocean in its westward course, the Celtic migration was apparently reacting eastward. As to the boundary between the Germans and the Celts at this early period, it cannot be clearly defined. Most probably it was formed by the Rhine, from its sources in Switzerland to its mouth in the North Sea. The later history of the Celts is well known, and we need not here concern ourselves with the numerous invasions, Roman, German, Saxon, and Norman, to which they were subjected, and by which they were crowded into their present contracted domain.

But there are phenomena of race-variation in the history of the Celts to which some allusion must be made. When they first appeared in history they were of the pure blond type, and had the stature, physical strength, and fierceness of the barbaric Xanthochroi. "The Gauls," says Ammianus Marcellinus, "are almost all tall of stature, very fair and red-haired, and horrible from the fierceness of their eyes; fond of strife and haughtily insolent."[1] This, in fact, seems to have been the character, physical and mental, of all the Aryans who peopled the north and west of Europe,

[1] Latham, Natural History of Man, p. 194.

though it is by no means the case with the great mass of the peoples who are supposed to be descended from them. There seems to have been a very considerable infusion of a darker and smaller human element, — probably that of the aborigines, who doubtless much exceeded their invaders in number. In this way a vigorous influx of Melanochroic blood seems to have entered the veins of the blue-eyed and fair-haired primitive Celts.

From this combination comes the French population of to-day. Here we find a blond type yet existing in the North, while the central districts are occupied by the modern Celtic type, with upturned nose, somewhat depressed at the bridge and but little projecting, hair brown or dark chestnut, eyes gray or light in shade. Such are the people of Auvergne and the Low Bretons, — a small and swarthy, round-headed race. In southern France several types are found, and there seems a strong infusion of Basque and Berber blood. Something similar might be said of the Celtic districts of the British Islands. In fact, as the Celts conquered the ancient inhabitants by force of arms and of energy, the aborigines seem to have conquered the Celts by force of numbers. As M. Roget says, the blue-eyed, fair-haired, long-headed Celt has been giving place in France in a direction from the south to the north to a more ancient, dark-eyed, black-haired, round-headed type. There has been a corresponding change in character, and the impulsive, emotional mentality of the aborigines has triumphed over the more staid and thoughtful character of the Xanthochroic man.

So far as indications go, the path of the Celts from ancient Arya was due westward through middle Europe. They seem to have been followed by two other Aryan branches, — that of the Teutons, which trod in the Celtic path, and that of the Greco-Italic section, which may have pushed through the mountains and along the southern shores of the Black Sea, making Asia Minor its line of march. Neither of these subsequent invasions found as easy a task as that of the Celts, if we may judge by indications. The latter had only the aborigines to deal with; but the former came into contact with the fierce and warlike Celts, who were quite their equal in vigor and in the arts of war. Perhaps in consequence of this we find a diversion in these later lines of march, the southern branch confining itself to the peninsulas of Greece and Italy, while the northern branch pushed into upper Germany and sent its leading tribes far into the Scandinavian peninsula. The Celts may have stood as a firm wedge in the median line of Europe, splitting the subsequent lines of march, and forcing them to diverge to the south and the north.

Of these migrants the Teutonic were strongly of the xanthous, or blond type, and their Scandinavian section has continued so to this day, preserving for us in considerable purity that type of physical and mental character which has been so greatly modified elsewhere by the infusion of alien blood. The intellect of this Xanthochroic division, as described by Dr. Knox,[1] is not inventive, has no genius for the abstract, no love for metaphysical speculation, cares nothing for the transcendental, and is naturally sceptical, bringing everything, even its religious faith, to the test of reason. In this description we seem to have the highest outcome of the practical Mongolian mind, — an intellectual condition capable of the greatest things when once kindled by the fire of imagination, but unprogressive in itself.

[1] The Races of Man, p. 344.

The ancient Aryan inhabitants of Germany are described by Tacitus as a tall and vigorous people, with long, fair hair and fierce blue eyes. They lacked somewhat the reckless impulsiveness of the Gauls, yet were as fierce and brave as the latter. To speak, however, of a Celtic followed by a Teutonic Aryan migration, is to deal with the subject from a general point of view. There seem to have been many successive waves of the Aryan flood, each pushing forward the preceding, and giving rise to numerous separate tribes. It is only linguistically that they can be called distinctively Celtic and Teutonic. They formed successive migrating sections of the two most northwesterly branches of the Aryan stock. Thus Cæsar describes Gaul as inhabited by three distinct nations, — the Aquitani, the Gauls, and the Belgæ. Of these the Aquitani are supposed to have been aborigines, with some Celtic admixture. The Gauls are described as bright, intelligent, vivacious, frank, open, and brave. The Belgæ were more staid, less active, more thoughtful, and less easily exalted or depressed. They approached the Germans in character, and had least varied from the primitive type. The Germans, in their turn, were divided into several branches which spoke distinct languages, and into numerous tribes. Probably they entered the country in several successive waves from the east. The Xanthochroic Germans of the time of Tacitus, however, have since then suffered much the same fate as the Celts. There has been a great amount of mixture with a dark-haired people, and the modern Germans have lost all distinctiveness of race, though they are less Melanochroic than the peoples of southern Europe. Probably they, like the Celts, amalgamated with their conquered subjects and with the Melanochroic peoples bordering their domain on the south. However that be, there is to-day no distinctive Teutonic type; every variety of man, from fair to dark, can be found on German soil.

Tacitus gives us much interesting information concerning the habits and conditions of the Germans of his time, which is of importance from its probable close affinity to the life of the primitive Aryans. Their dress seems to have been very scanty, consisting mainly of a mantle of coarse woollen stuff, flung over the shoulders and fastened with a pin or a thorn. Farther north mantles of fur were worn. Their dwellings were low circular huts made of rough timber, thatched with straw, and with a hole at the top for the escape of the smoke. The inner walls were roughly colored, and cattle sometimes shared the interior with the family. Their dwellings did not stand close together, but apart and scattered, each freeman choosing his own home. Their favorite occupations were war and the chase, and there is very little indication of agriculture. When not thus engaged, they often lay idly on the hearth, leaving all necessary labor to the women and to men not capable of bearing arms. In their social gatherings drunkenness and gambling were prevalent evils. Their arms were a long spear and a shield, with occasionally clubs and battle-axes. Each freeman was expected to bear arms and march to battle under his own clan head, the tribe being led by its hereditary chief or its chosen *herzog*, or general. Thus constituted, they rushed to battle, roused to fury by the excitement of war, and striving to intimidate their foes by loud shouts and the clashing of shields. The loss of a shield in battle was the loss of honor, and the despair of the loser frequently ended in suicide.

Latest of the northern Aryan migrations came that of

the Slavonic tribes, pushing hard on the heels of the Germans, and driving them forward into the heart of Europe. This movement was probably contemporaneous with the historic period of southern Europe. It carried the Slavic race much farther into Europe than it has been able to maintain itself, since the reaction of German valor has driven back the Slavs to their present borders, — the western limits of Poland, Bohemia, and Russia. In this connection it is somewhat singular that both Berlin and Vienna, the German capitals, stand on ancient Slavonic ground. More to the south they have held their own, — in eastern Austria and in the northern and western districts of European Turkey. Probably one of the earliest of the Slavonic movements was that of the Lithuanians, — a people with a language of distinct individuality, who have preserved the Xanthochroic physical character far better than their Russian kindred. Back of all these outlying branches came the Russians proper, — seemingly the last of the Aryans to leave their ancestral home. In fact, if our idea of the location of this home is correct, the Russians still occupied it at the opening of the historic period, or had moved but a short distance to the west. In the fifth and sixth centuries we first gain a clear vision of this people, then occupying a limited region in the territory of Little Russia, in the neighborhood of the present Russian district of Kiev. Here was the germ of the great empire which has since so widely spread, under rulers of Teutonic blood. The region indicated is in the immediate vicinity of that which we have considered to be the probable locality of the northern section of the primitive Aryans. The Slavonic branch was doubtless the last to leave the old Aryan home, if it can be said to have left it at all. There certainly remains a people of Slavonic affinity in the region which we have conjectured to be the mountain birthplace of the Aryan race; namely, the Ossetians of the Caucasian range. "This people," says Pallas, "exactly resemble the peasants in the north of Russia; they have in general, like them, either brown or light hair, occasionally also red beards. They appear to be very ancient inhabitants of these mountains." The Slavonian migration, after its first fierce outward push into western Europe, apparently became a very deliberate one. It is important to notice that it has not yet ceased. From the first entrance of the Slavic race into history it has been yielding to the pressure of the Teutonic race in the west, but pushing its way persistently to the north and east. At the same time it has been mingling intimately with the Mongolian race, and has acquired strong peculiarities of feature and character in consequence. The Mongolian blood and type of mind have partly reconquered the Russian from the Aryan race.

The Slavonic movement has been one of slow agricultural expansion rather than of warlike enterprise. The Slavs are the least restless, the least warlike, and the least progressive of all the Aryan branches. They have the most faithfully preserved to modern times the ancient institutions and the antique grammatical methods; and the indications are that they could have indulged but little in the disturbing game of war and migration in the prehistoric period. They seem to be the home-staying Aryans, the keepers of the old homestead, who remained on the ancestral domain while all their brethren went abroad. Their movement has been mainly that steady outgrowth of the farm before which the nomad horde can never sustain itself.

Gibbon remarks of them that "the same race of Sclavonians appears to have maintained, in every age, the possession of the same countries. . . . The fertility of the soil, rather than the labor of the natives, supplied the rustic plenty of the Sclavonians. Their sheep and horned cattle were large and numerous, and the fields which they sowed with millet or panic afforded, in the place of bread, a coarse and less nutritious food."[1] Such are the conditions which probably existed in the primitive Aryan home. The ancient Slavs were not distinguished for bravery. Their military achievements were, as Gibbon remarks, those of spies and stragglers rather than those of warriors, and they were incessantly exposed to the rapine of fiercer and more warlike neighbors. This hardly applies, however, to the southern Slavonians, who invaded the eastern Roman empire with vigor and success, and who treated their prisoners with the most savage cruelty.

The characteristics of the Russian Slavonic population, as above given, are not those of the Aryan race as generally known. In fact, the Slavs of Russia have lost their distinctive Aryan character yet more fully than the Celts have in the West. In both cases the language and institutions have been retained, but the race-distinction has largely vanished. The Russians frequently present a close resemblance to the Mongolian type, and either have become largely mingled with, or originally closely resembled, the Finns, as is indicated by the dark skin and yellow beard so common among the peasants. The face is hollowed out, as it were, between the projecting brow and chin. The race is tall, but not robust, strong, but not energetic, and displays a general character of apathy.

[1] Decline and Fall of the Roman Empire, iv. 197.

They lack invention, but are admirable imitators, like the Mongolians. In fact they present decided Mongolian characteristics. In the southeast the Slavs are dark, with dark hair and eyes. These comprise the Croats, the Servians, and the Slavonians proper. But the Slovaks of Austria possess the fair skin and red or flaxen hair of the northern Russians. It is, in truth, a race of manifold mixture, the only character common to all Slavs being brachycephaly, — a Mongolian characteristic. It is a race which lacks much of the intellectual vigor and the restless energy of the purer Aryans. These remarks, however, apply mainly to the peasantry. In the blood of the ruling class there is a considerable infusion of the German and Scandinavian element, and it is to this class that we owe the migratory activity of modern Russia. The characteristic of the peasantry is apathetically to stay where they are placed, though always ready to migrate where a decided agricultural advantage appears. This survival of an antique custom is a valuable aid to the colonizing enterprise of the Government.

The movements of the northern Aryans were matched by an equally active expansion of the darker-skinned southern sections, the fathers of the Greek and Latin, the Persian and Indian, civilizations. We know as little concerning the dates of these movements as of those of the North. In speaking of the Celtic as the earliest migration, this may apply only to the northern movement. That of the South may have been contemporaneous with or antecedent to it. When history opens, the Celts are still in active movement. They have not completed their work. The Germans are visibly moving, and the Slavonic tribes have probably not yet left the region of ancient Arya. But no historic trace of such a movement can be found in the story of the Greeks and Italians. When first seen they are in full possession of their historic realm, and retain not even a tradition of a migratory movement. They proudly term themselves autochthones, the original possessors of the soil. We can deem their movement as contemporaneous with, or later than, that of the Celts only from its southward diversion and the fact of the Celtic possession of central and western Europe. Yet this may be due to the one migration being to the north, and the other to the south, of the Black Sea.

In our scheme of the primeval Aryan home the ancestors of the Greeks and Italians occupy the southwestern region, — perhaps continuous in their northern borders with the Celts, if we may judge from certain affinities of language. Their location is the Caucasian mountain district and the northeastern region of Asia Minor. Such seems probable from what we are able to discover of their movements, and also from their much greater loss of the Xanthochroic race-element than in the northern Aryans. Though not destitute of the blond type of complexion, the brown type was the prevalent one. They had probably considerably mixed with the brown Southerners before their migration; yet they never forgot that the blue-eyed and fair-haired type was that of their ancestral race, and to the last they preserved an admiration for it.

The line of Grecian march, so far as we can trace it by linguistic evidence, appears to have been through Asia Minor. The Greek testimony would make Greece their native home, and the settlements in Asia Minor the outcome of colonizing movements. But modern research has led to a different opinion, and indicates that at least the

Ionians originally came from Asia Minor. The typical Hellenes can be traced, with considerable assurance, to the highlands of Phrygia, — a fertile region of northwestern Asia Minor, such as a tribe of mountaineers would naturally make a stopping-place in its westward march. Here perhaps they long halted, increased greatly in numbers, and gave off successive divisions, which pushed westward into Greece, while the vanguard of the march made its way into Italy.

All we know of the history of early Greece is that it was inhabited by a people called Pelasgians by the later inhabitants, but of whose derivation we are in absolute ignorance. Much has been written about them. We are told of a great wave of migration which carried over the Hellespont into Europe a population which diffused itself through Greece and the Peloponnesus, as well as over the coasts and islands of the Archipelago. To this antique Aryan tribe are ascribed the most ancient architectural monuments of Greece. We are further told that the coming of later tribes pushed forward this Pelasgian outpost until it overflowed into Italy, while it vanished from Greece either by destruction or amalgamation. This, however, is all pure conjecture; it has no historic basis. We know nothing of the origin, race-character, or degree of culture of the early inhabitants of Greece, though there can be little doubt that the Aryans made their way by successive waves into Greece and Italy.

Before the final Hellenic migration began, the Hellenes had apparently divided into two distinct sections, well marked in language and character, — the Doric and the Ionic. A third section, the Æolic, separated at a later period. It is conjectured that the Dorians continued to occupy the highland region, while the Ionians moved south to the sea-coast of Asia Minor, where they found a softer climate and gained new habits of life. This conjecture seems borne out by their subsequent character and history. Our first historic trace of the Dorians is in the highlands of Macedonia. Here they displayed the type of the hardy mountaineer, which was probably original with them. From this position, at a later date, they pushed southward and occupied the Peloponnesus, their historic home, forcing back the Ionians who had preceded them.

We can recover no historic trace of the primitive Ionians. They probably made their way into Greece over the islands of the Archipelago, having long before come into contact with the Phœnician navigators and gained the germ of the maritime skill and enterprise which were afterwards to distinguish them. Spreading themselves over these numerous and fertile islands, they finally entered Attica, the famous centre of their future civilization. But it is highly probable that they still held possession of the coast of Asia Minor, and that what were afterwards described as colonies were really the original Ionian settlements. Here, at least, their civilization first budded. Here the Grecian arts first grew into prominence. Here was the land of the Homeric song and the scene of the great poet's life. Hence came the earliest song-writers, philosophers, and historians to the rising commercial city of Athens, to gain in its rich precincts the reward of their genius and to implant that seed of thought which was afterwards richly to grow and bloom on Attic soil. That later colonies, Doric, Ionic, and Æolic, settled on the shores of Asia Minor, there is historic evidence; but they evidently settled among Greeks, and found there in a developing condition that literary and artistic culture which was afterwards to gain its highest expression on the peninsula of Greece.

As to when and how the Aryans came into Italy we know absolutely nothing. We find them there at the opening of history, and that is all. The earliest Greek colonies in the south of Italy met there two peoples, called by them the Iapygians and the Ænotrians, whom they looked upon as Pelasgians or as remnants of the most ancient known population of Greece. They were possibly Aryans, but of this we cannot be sure; the extant relics of their language are too slight to be of much utility. Central Italy was occupied by numerous tribes, which have been divided into five groups, — the Umbrians, Sabines, Latins, Volscians, and Oscans. There is good reason to believe that these were all of Aryan stock. The Umbrians have left an important linguistic record in the celebrated inscriptions known as the "Eugubine Tablets," which indicate a very primitive Aryan dialect and stamp the Umbrians as one of the most ancient Aryan nations of Italy. As for the remainder of Italy, the North was occupied by several distinct peoples, prominent among them being the strong Celtic settlement known as Cisalpine Gaul. Southward lay the land of Etruria, occupied by the remarkable people who rose into the earliest Italian civilization, but whose ethnic affinities are still a puzzle. Whether they were or were not Aryans is a question that remains to be settled. All we positively know is that ancient authors represent them as a people wholly distinct from all others in Italy. As for the Latins, the race that was subsequently to make such a remarkable figure in the world, and so greatly to advance the Aryan civilization, their origin is in great obscurity. Their earliest traceable home seems to be the central Apennines, and their language has a considerable infusion of the old Greek element, which indicates a very ancient branching off from the original stock of Greco-Italic speech.

We have one remaining Aryan migration to trace, — the Indo-Iranic, that which carried the fathers of the Hindu and Persian empires to their temporary Bactrian home. This branch of the Aryan stock, in our scheme of the ancient home of the race, would have its location in the southeastern Caucasian region, impinging on the southern shores of the Caspian. Here, like their neighbors to the west, they seem to have largely lost the distinctive Xanthochroic type, and to have been greatly modified by an infusion of the Melanochroic element. Their migration may have been considerably later than that of the Greeks. Quite possibly, indeed, an Iranian pressure may have instigated the Grecian movement, if we may judge from the fact that Armenia is to-day occupied by an Aryan people who speak an Iranic dialect. As for the march of this branch of the race, we have no more historic evidence than in the case of the other branches. All we can discover is an extended line of Aryan peoples, leading from the Ossetes, who occupy the pass of the Caucasus, successively to the Armenians, the Kurds, the people of ancient Media and Persia, the Afghan and Belooch Aryan tribes, and the Hindus of the Indus and Ganges. At every point on the long line of march divisions of the migrating army were seemingly dropped, or perhaps the expansion of a growing people pushed its vanguard farther and farther over the eastward path, on a route probably much easier than that leading to the civilized regions of the South.

Of all this, however, we have no historic evidence. Though we are now dealing with a people who possess a considerable literature, dating from a period when their migratory movement was yet far from completion, yet this literature is the reverse of historical. It is simply calculated to bewilder and lead astray the earnest students of history. The Vedas of the Hindus, indeed, make no pretence to be historical. The Zend-Avesta of the Persians, while not historical, lays down a geographical scheme, which forms the sole basis for the selection of Bactria as the primitive Aryan home. Yet this Avestan geography is of the most mythical and unsatisfactory character. In the "Vendidad" are enumerated sixteen lands created by Ahura Mazda. Many attempts have been made to identify these, and draw historical conclusions from their order in illustration of the line of Iranian migration. These efforts have proved signally unsuccessful. Several of the lands named are clearly mythical, and of only nine can the location be traced. Yet in naming these the Persian author seems to have wandered at random over the map, without regard to the cardinal points. No conclusion can be drawn from their order of succession, since they have no order.

This geographical record, however, appears to indicate the region of ancient Bactria as the point of common residence of the Hindus and Iranians ere yet they had divided into two sub-branches and begun their final migration. It was a land adapted to their needs, with its mountain-slopes, its tracts of rich soil and fine pasture-land, its abundance of oxen and horses, its warm summer airs on the northwest terraces of the Hindu-Kush. But that it formed the original Aryan home there is not a shred of evidence, while such an idea is surrounded by insuperable difficulties. In all probability it was the halting-ground of the vanguard of the Aryan march to the East, a land in which they may have long rested, and where their numbers may have greatly increased.[1] All we really know is that, after probably a long residence in this locality, during which the primitive Aryan ideas became much modified, a division took place. Some claim that this was a religious schism. Of this we have no evidence other than the strong religious fervor manifested in their literature, and the diversity of opinion concerning the gods that appears in the most ancient documents of the Hindus and Persians. It is assumed that a group of sectaries, under the leadership of Xarathustra or Zoroaster, broke off from the main stock and made their way towards the highlands of Iran, retracing, as we assume, their original path, probably long forgotten. Here they established themselves, developed the distinctive Zoroastrian faith, and became the root-bed of the future great empire of Persia.

There is nothing surprising in such a reverse movement. The whole of the Aryan population of Bactria seemed to be in motion, and expanding in all available directions. The Indic branch was pushing toward the rich plains of the South, and there was but one path left open for the Iranic, — that leading to the Persian highlands. The march of the fathers of the Hindu race can be traced with some clearness. They seem to have pushed out from

[1] A study of the map of Asia shows a comparatively short route, by way of the southern shores of the Caspian, from the region of the Caucasus to that of the Hindu-Kush. It may be conjectured that the original Aryan migrants were forced to pursue this route by the hostile resistance to invasion of the primitive mountaineers of Persia, and that only after they had greatly increased in numbers and warlike strength in Bactria were they able to return and to cope with the foes whom they had avoided in their original march.

the western borders of Iran and made their way by slow stages and in successive tribes into the rich, warm, and moist valley of the Indus, seeking a new home in these fertile plains. We can almost see them, in the pages of the Vedas, marching resolutely south, singing their stirring hymns of praise and invocation to their deities, led by their priestly chiefs, and calling down the vengeance of the gods on their enemies, the *Dasyus*, the " raw-eaters," the " godless," the " gross feeders on flesh," the " disturbers of sacrifices," the " monsters " and " demons " who dared resist the arms of the god-sent, the Arya, the noble and ruling race.

This movement was in no proper sense a migration. It was, as we conceive was the case with all the Aryan movements, an expansion caused by increasing numbers and aided by hostile pressure from the rear. There are no signs of a march in force, but rather of the movement of successive tribes, each pushing the preceding one forward, and the whole slowly gaining possession of the broad region of the " five rivers," and extending to the great plain of the Ganges. We can trace the line of march in the Vedic hymns. The earliest ones disclose the Hindu tribes to the north of the Khyber Pass, in Cabul. The later ones were written and sung on the banks of the Ganges. Along the base of the Himalayas they pushed, and far down into that fertile and enervating land, driving the dark-skinned aborigines everywhere before them into the mountains and the jungles, and probably, despite their religious distaste, mingling their noble blood to some extent with that of these despised aborigines.

How long ago this was, can be conjectured with some degree of probability. The first occupation of the valley of the Indus, with its five tributaries, has been estimated, from what we know of the subsequent history of the Hindus, to have taken place about 2000 B. C. It could hardly have been more recent, yet it may have been more remote. According to the list of Babylonian dynasties given by Berosus, the western part of Persia was occupied by Aryans as early as 2500 B. C. All such estimates, however, must be taken with many grains of allowance.[1]

As to the physical and mental character of these eastern Aryans, something may be said. The Hindu type is decidedly Melanochroic. The Brahmin of the Ganges is marked by a high, well-developed forehead, oval face, horizontal eyes, a projecting nose, slightly thick at its extremity, but with delicately shaped nostrils, a fair but readily bronzed skin, and abundant black hair. Farther south the mixture with the aborigines has been so great that it is not easy to trace the typical Aryan. In fact there has never been a Hindu conquest of the southern half of India. There the Dravidian population still exists to the number of fifty millions, though all race-purity has vanished through the abundant mingling of types that has seemingly taken place. The mentality of the ancient Hindus was such as we might deduce from this mixture of blood, one with highly acute powers of reasoning, but

[1] This possibility of limiting the era of the Hindu-Iranian movement within historic times, in connection with the remotely prehistoric character of the early European movements, is a strong argument against the Bactrian locality for ancient Arya. No one can be asked to believe that Aryan enterprise began with difficult and distant migrations, and left the rich valleys of India, within easy reach, for its latest field of action. Such a reversal of the order of nature is inconceivable, and the probability is that the invasion of India was the final stage in a long-continued eastward migratory movement.

with perhaps the most developed and exuberant imagination that has ever appeared upon the face of the earth.

The Iranian populations of to-day — the Kurds, the Armenians, and the Tadjicks of Persia — are marked by black eyes and brows. The Tadjicks, the purest descendants of the old Persians, are described as of oval face, broad, high forehead, large eyes, black eyebrows, straight, prominent nose, large mouth, thin lips, complexion fair and rosy, hair straight and black, beard and mustache black and plentiful, and abundant hair over the whole body. In Afghanistan the pure Aryan type is frequently found. The Patans, or Afghan soldiers, are commonly brown like the Iranians, but many of them have red hair and blue eyes, with a florid complexion. This is particularly the case with the Siah Posh of Kaffiristan, a tribe which speaks a dialect derived from the Sanscrit. Thus in the Iranian branch of the eastern Aryans the Xanthochroic character has been much more fully preserved than with the Hindus. It is possible that the separation of the combined race may have been due to ethnic rather than to religious causes. The Iranians are highlanders to-day, and may always have been so. They may represent the mountaineer section of the original migrating horde, and therefore the one that had originally least of the Melanochroic element. Possibly they occupied in Bactria the highland region, and the Hindus the lower districts. If such were the case, we should have an additional reason for the Iranian movement towards the Persian highlands, and that of the Hindus towards the Indian plains. It is a case parallel to that of the Doric and Ionic peoples of Greece. In ancient Arya the Dorian and Iranian tribes may have been mountaineers, the Ionian and Hindu tribes lowlanders, and each may have been governed by this original habit in all subsequent movements. The Persians are distinguished from the Hindus by characteristics not unlike those separating the Dorians from the Ionians. They have the mental character of mountaineers, are brave, enterprising, earnest, and truthful, with a strong love of liberty, and much warlike energy. They lack the highly active imagination of the Hindus, but have a sound common-sense and vigor of thought which make them essentially practical in their religious systems. The Persian myths have had a profound influence over the practical religious history of mankind, while the Hindu belief forms the basis of all the involved figments of metaphysical philosophy.

But one thing more need here be said. Despite their many differences, there is a remarkable degree of homogeneity among the early conditions of the several branches of the Aryans, — alike in language, in religion, in political and social institutions, and in physical and mental character. This indicates an original great uniformity, a state of stagnant barbarity of long continuance, during which the Aryans greatly extended the borders of their primitive home without changing in any important degree their primitive institutions. For the second stage of progress a breaking-up and widespread migration were requisite, — contact with alien peoples, war, life in new lands, ethnic minglings, and all the varied influences which play upon an actively moving people, but to which a settled population is not exposed. To this diversity of influences, together with the inspiration of the old civilizations with which the outspreading race came into contact, we owe the highly developed Aryan enlightenment of the present age.

Briefly to summarize some of the conclusions of this chapter, it may be affirmed that the original Aryan migration had the character of an agricultural outpush similar to that which exists in Russia to-day. It was the natural expansion of an increasing race, at first of small, but of gradually growing enterprise, spreading from a central region in all directions to which fertility of soil invited. It was the onward step from farm to farm, with hostile aggression where this became necessary, the forward movement occasionally accelerated by a hostile push of other Aryan tribes from behind. These movements took place to all parts of the compass except that leading to the desert regions of Asia, and the whole intermediate region continued in Aryan hands. In their advance through Europe the Aryans have loosed their hold on no land which they once occupied, except where forced to do so by the invasions of the Huns and the Turks. In the East they have left communities in Armenia, Kurdistan, and other districts on their line of march, while the Aryan tribe of the Caucasus known as the Iron or Ossetes significantly occupies the path by which these southward movements must have taken place, — the Gorge of Dariel, the only natural road through the great mountain-chain. This tribe seems to have been left behind as the rear-guard of the Aryan army on its march to empire, while the Caucasus generally has been occupied by alien peoples.

It was only at a later period, when migration and war had consolidated and given new energy and enterprise to the Aryans, that they ventured on bolder movements. We can perceive the gradual growth of this enterprise and power of warlike massing in the German tribes, to whom the immense wealth of Rome offered the strongest incitement to hostile aggression. Yet at no time did they make movements *en masse* like those of the nomadic Hunnish invaders. While crossing the borders into the Roman Empire, they held on persistently to their fields and forests at home.

The Aryan migration was evidently followed by an extensive intermarriage with the original inhabitants of the conquered territories. There is no evidence to the contrary, except in the case of the settlers in Scandinavia, who may have felt a strong antipathy to the widely different Lapps. Elsewhere, however, they found their new possessions occupied by tribes of Melanochroic blood, to whom the Xanthochroi have never shown any antipathy. Instead of annihilating or dispossessing these, they apparently simply subjugated them, and later on freely intermarried with them. Only thus can we understand the great change in physical characteristics of the Celts and Germans within the last eighteen centuries. In the former case the conquered must have much exceeded the conquerors in number, to judge from the strongly declared Melanochroic character of the modern Celts. As regards the Greeks and Latins, the Hindus and Persians, it is quite probable, as we have already conjectured, that they had gained a strong infusion of Melanochroic blood before their migration. This was undoubtedly largely added to after reaching their new homes, and particularly so in the case of the Hindus, who must have been greatly outnumbered by the aborigines of their conquered territory.

Yet in all these cases the Aryan type of language held its own persistently, doubtless adopting many words from the dialects of the conquered races, but vigorously maintaining its structure, and forcing out all the aboriginal tongues. This indicates that the aborigines were in every instance subordinated to the conquerors, who retained their ascendency firmly during the subsequent period of amalgamation. Of variations of linguistic structure the most marked were those which took place in the Celtic dialects, which seem to have had impressed upon them some of the characteristics of the aboriginal tongues, yet not sufficiently so greatly to affect their Aryan type.

IV.

THE ARYANS AT HOME.

WHAT can we know about the mode of life of a group of barbarians who have become extinct as a primitive community without leaving a trace of their existence upon the face of the earth, who have written no books, carved no monuments, built no great works of architecture? The early Chinese and Egyptians, probably their contemporaries, have left abundant monuments, — written, carved, erected, and excavated; but the Aryans ate, drank, fought, lived, and died without a thought that the world to come might be curious about their doings, and without an effort to stamp in stone, brick, or earth the story of their existence. They had not yet reached that stage of development in which men begin to think they are doing great things and living great lives, and become anxious to astonish the future world with a knowledge of their prowess. This wish to astound posterity is a feature of one stage of every advancing civilization. Primitive barbarism troubles itself but little about the curiosity of the future. High civilization is more concerned in working for the needs of the present. But the intermediate stage of budding civilization has always wasted its strength in building great tombs, pyramids, temples, and the like, as monuments of its greatness, toiling with the strength and blindness of the Cyclops to leave a message of empty wonder for the world to come.

The antique Aryans had not reached this stage of development. And yet they have, without knowledge or intention, left a record of their lives and institutions but little less complete than that of their fame-seeking civilized contemporaries. The political relations of the modern world are the growth of the seed which they planted. The religions of the mythological age were the unfoldment of their germ of faith and worship. The languages of modern times are full of words which this antique group spoke in their primeval homes. All these lines of development have become great trees; but they can be traced back to their roots, and in these roots we possess the life-conditions of our ancestral clan.

As we have already said, all the languages of modern Europe, the English, the Romanic, the German, the Celtic, the Slavonic, and the Lithuanian; those of ancient Europe, the Greek, the Latin, the Teutonic; those of southern Asia, the Sanscrit, the Persian, and their several minor dialects, — are not alone closely similar in grammatical structure, in skeletal type, as it were, but also are full of verbal affinities. From Ireland on the west to India on the east we find words essentially the same used to designate the same things. Very many such words exist, — far too many to suppose that these languages could have gained them by borrowing from one another. And these words are not the terms employed by civilization to designate its newly acquired treasures, but they are the names of things and ideas of simpler and more antique character, the titles of the possessions and conditions of barbaric life, for which every nation, if it had no primitive names, would have been forced in the early stage of its existence to invent names for itself. The conception, therefore, that these common terms were acquired during the process of national development by borrowing or, like articles of commerce, by interchange, cannot be entertained for a moment. But if this explanation be thrown aside as inadequate, there remains only that of a common origin. We are forced, in fact, to believe that all these widely separated nations are descendants of a single primitive people who once occupied a single, limited area, from which they have outspread over the earth, and who spoke a single and simple language, from which have come the complex and varied systems of Aryan speech.

We have already sought to trace the origin, the primitive locality, and the early migrations of this people. A yet more interesting inquiry is before us, — that of their mode of life. What did they know; how did they live; what was the character of their possessions? — such are the queries which we must now seek to answer. We look back far into the darkness of the past as into a mist-shrouded valley, and perceive at first only impenetrable gloom. But finally a ray of light of growing strength makes its way through the thinning vapor, and by degrees a broad scene of busy life is revealed to our eyes, — not with much clearness, it is true; not without wisps of shadow clinging to and half enveloping its objects; yet sufficiently clear to yield a very considerable knowledge of the conditions of that long-clouded scene of ancient life. This revealing ray has sprung from several sources, one of the most important of which is that of comparative philology.

In isolating the words common to the Aryan languages, it has been necessary to place them in two divisions. One is of words common to a part only of these languages; the other of words common to the whole. The former series indicates that certain branches of the Aryan race, after breaking off from the main stem, again divided after their special dialect had made considerable progress. Such was the case with the eastern branch, and thus we may account for common words in the Indian and Iranian tongues which do not extend to the other branches of the race. This special community between the languages of the two great divisions of the eastern branch is paralleled by similar special resemblances in the west, as between the Greek and Latin. Efforts have been made, in consequence, to divide the Aryan race up into secondary, or sub-races, the product of a primary division, each of which sub-races made considerable progress before a new division took place. But from these efforts no very satisfactory result has been achieved. Several unlike schemes have been proposed, each of which has been contested and denied. We need, therefore, concern ourselves here only with the original Aryans, without heed to their assumed but as yet unproved sub-branches.

The persevering and critical labor of the students of language has, as we have said, isolated numerous words which must have been in use by the Aryan family before its separation, since they are still in use by all, or nearly all, its descendants. This work has gone so far that we have now a dictionary of the ancient Aryan in three stout octavo volumes.[1] And August Sleicher has taken the trouble to write a short story in this prehistoric language. It is quite likely, indeed, that the ancestral Aryans would have had some difficulty in reading it, since it cannot be supposed that the exact form of any of their words has been preserved; yet it is curious, as showing the great

[1] Fick's Comparative Dictionary of Indo-Germanic Speech, 1874–76.

progress which has been made during a few decades of persistent study.

Words indicate things and conditions. No people has ever invented a vocal sound without the purpose of naming something which they had or knew. It cannot be supposed, however, that the Aryan words conveyed to the minds of their early speakers the exact meaning which they do to ours. The words of our languages have become as full of mental as of physical significance. Philosophical conceptions spread like a network through the substance of our speech. But we have now to deal with a people who had not devised a philosophy and had little conception of mentality. They knew what they saw. They named what their eyes beheld or their hands encountered. Their world existed outside them. The vast world of the mind was as yet scarcely born. Numerous evidences of this might be quoted. The names of the family relations, for instance, originated in physical conceptions. The Sanscrit *pitar*, "father," comes from *pa*, "to protect." The original meaning of *bhratar*, "brother," was "he who carries or assists." *Svasar*, "sister," signified "she who pleases." *Duhitar*, "daughter," is derived from *duh*, a root which in Sanscrit means "to milk." The daughter of the primeval household was valued mainly for her use as a milkmaid. Thus what seem to us the most primitive of words were really derived from preceding physical terms. As yet no general or abstract conceptions existed. Indeed we may come to far more recent times without much improvement in this respect. Old Anglo-Saxon, for instance, is far richer than old Aryan. Yet if we should seek to converse on philosophy or science in Anglo-Saxon speech we should soon find ourselves in difficulty. Only by a free use of metaphor, and mental applications of words which have only a material significance, could any progress be made in such a task. It is very probable, however, that the antique Aryans had long forgotten the derivation of their words; they were mere technical symbols to them as to us. Their language had been developed probably many long centuries before the era of their dispersal, and linguistic decay had already set in. We know far more than they did of the origin of their words, from our method of isolating the roots of language, and reaching down to the deepest-buried seeds of meaning.

Let us seek to rehabilitate this ancient Aryan community, so far as our knowledge of their words enables us to do so. For this purpose we shall mainly follow Professor Sayce[1] in his graphic rebuilding of old Arya from the words given in Fick's "Comparative Grammar." If we look far back through the revealing glass of science we seem to behold these active aborigines on their native plains engaged in all the vocations of a simple life. We see them employed in a twofold duty, — that of pastoral, and that of agricultural life. Abundant flocks are scattered over their grassy commons attended by the diligent herdsman. Of domesticated animals the cow was their most valued possession, as it still is with the pastoral tribes of northern Asia. But in addition they had the horse, the sheep, the goat, and the pig. There is nothing to show that the horse was ridden. If we judge alone from the indications of language, we must believe that it was, in common with the ox, used only for drawing. Nor is there anything to show that the dog was known in other than its wild state.

[1] Introduction to the Science of Language. A. H. Sayce.

And yet the exigencies of pastoral life may have required the modern use of these animals. To their sheep and cattle pastures the Aryan herdsmen added the shelter of stables, sheepcots, and pigsties. Of other domesticated animals may be mentioned the goose and fowl as probable, while the bee was undoubtedly one of their valued possessions, its honey being made into mead, — then and long afterwards a favorite Aryan beverage. Their chief ordinary drink, however, was the milk of the cow, sheep, and goat; and the morning milking scene by the daughters of the tribe doubtless closely resembled that still seen on the Asiatic steppes among the pastoral nomads of that region.

The community with which we have at present to deal was not a nomadic one. It had doubtless passed through that stage of existence; but at the time in which we behold it the development of agriculture had tied it to a fixed locality, and the interests of agriculture were steadily rising into prominence. There are indications to show that in the early days of the development of Aryan speech the pastoral interests were largely in the ascendent. But at the period immediately preceding the Aryan dispersal, agriculture had become considerably developed, the tribes were settled in definitely arranged communities on a fertile region, well watered and wooded, and farming and herding had become common industries of the people, without the wide division between these interests which we now find in the desert regions of Arabia and Turkestan, with their fertile oases alternated with scanty pasture regions.

The antique language has abundant indications of such a primitive supremacy of pastoral interests. The names for many of the family and tribal relations, for property, trade, etc., for inn, guest, master, and king, were taken from words that applied to the herd. Dawn signified the mustering-time of the cows. Evening was the time of bringing home the herds. In the word "cow" itself we have "the slow walker;" in ox, "the vigorous one;" in dog, "speed;" in wolf, "destroyer," etc. All this indicates that the era of development of the language was an era when pastoral interests were very prominent in men's minds.

But evidently at the period of the Aryan dispersion the interests of agriculture were becoming dominant, and those of a pastoral life secondary. We have warrant for this in the plentiful survival of common agricultural terms, and in the word by which the eastern Aryan migrants called themselves at their first appearance on the stage of history, — *Aryas* in the Vedas, *Airyas* in the Zend literature, — and from which their modern title has been derived. This word comes from a root which signifies "ploughing." It grew eventually to mean "honorable," or "noble." The Aryans, not without warrant, considered themselves the noblest of human races.

If we now turn our mental gaze from the pastures to the farming lands we see indications of a different mode of activity. Here the earth is being turned up with a rude plough drawn by the slow moving ox, or possibly the horse. There the hay is being cut with the sickle. Yonder are fields of ripe and waving grain of at least two kinds. Just what grains these were, we cannot be quite sure. One of them seems to have been barley, — the cereal of cold climates. The other may have been wheat, though this is far from certain. These, with a few garden vegetables, are all we can perceive through our highly imperfect observing-glass. We can, however, see wheeled vehicles of some sort, drawn by yoked oxen, and bringing the harvests from the field. We can likewise perceive these antique farmers threshing and winnowing their grain and grinding it in mills. We have their words for wagon, wheel, and axle, and also for hammer, anvil, and forge, — the latter showing that the smith was an active member of the community.

In the woods around them grew the pine and the birch, — trees of cold regions; and probably the beech and the oak, though this is not positive. As to what fruit-trees they possessed, we are in doubt; nor are we certain as to their knowledge of the grape. They appear to have had three metals, — gold, silver, and bronze. Their possession of iron, copper, and lead is more doubtful, and there is reason to believe that stone tools were still used. In fact, when we consider that metals may have been articles of commerce at an early date, and their names have travelled with them, the existence of common Aryan names for any metal is not as sure evidence of its early possession as in the case of many other articles, and it is possible that their actual acquaintance with metals was very slight. There is reason to believe, however, that the class of smiths was held in high honor, and that they sometimes had supernatural powers attributed to them, as among other barbarian communities.

The people whose life in the dim depths of time we are thus observing had left behind them the tent-stage of existence. They dwelt in houses of wood, with regular doors, instead of the hole through which the tenants of many northern habitations crawl. We cannot identify any window. Straw seems to have been used to thatch the roofs. It is possible that these houses were but rude huts. They were combined into villages, whose name still survives in the *wich* or *wick* now often used as a termination of the names of towns. There seems also to have been a fortress, with protecting wall or rampart.

As for domestic life and comforts, we know that baked pottery was in common use, formed into vases, jars, pots, and cups, some with the ends pointed so as to be driven into the ground. This pottery may have been ornamented by painting in colors. Vessels of wood and leather were also probably in use. The hours of relaxation seem to have been softened by music, derived from some stringed instrument. The food used appears to have included baked or roasted meat, and the eaters of raw flesh were looked upon as utter barbarians. Quails and ducks were eaten, and a black broth was apparently a principal article of food. Their meal was baked into bread, and apples may have been one of their edible fruits. Salt was used as a condiment. Quite likely their diet was considerably more varied than this, since many names of articles of food may have died out of use, or been replaced by others in the long course of time. Of the other household treasures may be mentioned *makshi*, "the buzzer," our common fly. With him was associated the less desirable flea, while the prowling mouse made up a trio of domestic pests. The art of medicine was as yet in embryo, but our ancestral clan was by no means free from the ravages of disease. Two names of diseases have survived, — consumption and tetter. As for cure, the power of charms seems to have been mainly relied on.

In these households strict monogamy prevailed. There was but one husband and one wife, and the family relations were clearly defined. In addition to words for father, mother, son, daughter, brother, sister, etc., they had separate words for a wife's sister, *syâlî*, and for a brother's wife, *yâturas*. The father was lord of the household, and the wife its mistress; the subordination of the younger members of the family to parental authority being far greater than in our era. The names of these antique Aryans were composed of two words, as now. We may instance *Deva 'sruta*, "heard by God," as the title of one of our extinct ancestors. As for their domestic industries, they seem to have possessed the arts of sewing and spinning. Wool was shorn and woven, and linen was known, though probably little used. The art of tanning was practised, and leather was much used for clothing and other purposes. Their dress apparently consisted of a tunic, coat, collar, and sandals, made of leather or of woven and sewn wool. But if we may judge from what we know of the early Germans, Slavs, and Celts, they were not greatly protected by clothing from the cold.

If now we leave the domestic and industrial conditions of the Aryans, and seek to follow them in the more stirring details of their active lives, we behold them engaged in what to them were doubtless nobler pursuits. Here we perceive our ancestor actively engaged in the chase and daringly entering into combat with the savage bear and wolf. Of smaller game he seems to have pursued the hare, beaver, and badger, and probably the fox. The wild duck was one of his game-birds, and he knew several other birds, such as the vulture, raven, starling, and goose. He had the custom, preserved till a much later period, of divining the future from the flight of birds, particularly of the falcon. The serpent was known, and probably both hated and revered for its deadly and mysterious power. Of his water-dwelling game we may name the otter and the eel, the crab and the mussel. But his knowledge of fish must have been very limited if we take language for our guide.

Changing our field of observation, we behold him boldly embarking on the waves of the great salt lake which adjoined his native land. The name he gave this watery expanse is still preserved in *meer*, — a word which has been since applied alike to sea and lake, moor and morass. Here he launched his boat, guided it by a rudder, and propelled it by means of oars. His barbaric intellect was not yet equal to the device of the sail, — or at least he has left no word to signify that he had learned to spread the broad sheet to the winds, and by their aid to avoid the laborious straining of the muscles.

A glance in still another direction shows him to us engaged in what he probably considered the noble pastime of war. That he was of belligerent disposition we have every reason to believe, judging from the irascible temper he has transmitted to his descendants; and doubtless his peaceful labors were frequently broken by warlike raids upon neighboring tribes or by fierce defence of his home and fields against hostile invaders. In this stirring duty the axe was apparently his chief weapon; but he fought also with the club and the sword, while he wore the helmet and the buckler for defensive armor. The bow was also probably one of his implements of offence. With these weapons the blue-eyed and stout-hearted champion doubtless fought sturdily for home and freedom, or for fame and spoil, doing doughty deeds of valor which may have roused to noble inspiration the minstrels of his tribe, yet which have vanished in the night of time and thrown not a ray of their lustre down to our remote age. As yet no Homer had arisen to make imperishable the deeds of warlike glory.

As for the acquirements of this strong-limbed and active barbarian, beyond the requisites of industry and war we know very little. He was acquainted with the decimal system of numeration, counting by fives and tens, with his fingers and toes as guides, at least up to a hundred. The year was divided into lunar months, the moon being to him the measurer of time. He doubtless had abundant superstitions. The evil spirits of night and darkness pursued and affrighted his shrinking soul. Their symbol to him was the serpent. Night was the demon, *aj-dahaku*, the biting snake. Then was strongly felt the consciousness of sin, when the gloom of midnight had densely gathered, and ghosts and witches held high festival in the air. But with the upspringing of the cheerful sun, and the forth-flowing of its gleaming rays over the earth's surface, these forms of terror shrank cowering to their dens and caves, and the Aryan stepped forth again in the proud consciousness of strength and valor, fearing nothing living or dead, and ready to cope with all the forces of the universe.

From such terrors and such deliverance, from the alternation of day and night, of summer and winter, arose his simple system of religious views. He worshipped the objects and the phenomena of Nature, and particularly the dawn and the other bright powers of the day. The broad blue sky was his supreme deity, to whom the stars and the moon were sons and daughters. To these he prayed and addressed his hymns, — the seeds of the complex mythologies into which his simple beliefs were destined to unfold. Of the many gods devised, he probably thought of and prayed to but one at a time; and supreme over them all was the mighty *dyaush-pitar*, the father of heaven, the guide and ruler of the universe.

We shall say as little here of his political as of his religious system, since we must deal with these more fully in future sections. It will suffice to observe that the family was the germ of the village community, which was constituted on the model of the household, and governed by the *vispati*, or head of the clan, or by the clan council. Over the larger political group ruled an elected chief of the tribe, who was assisted in his duties by a court or council, composed of *pataras*, or fathers of families. The landed property was held in common, the only individual property being the house, its court, its goods, and its cattle. The houses were grouped into villages, but the chief seems to have had his special residence and domain marked off from the common property. Each such community formed part of a larger group, — a township, to use a modern name. The separate townships were connected by roads, on which pedlers travelled with their wares. These communities had their laws, mainly the growth of ancient custom, for the prevention or punishment of crime. Justice was *aiva*, the path of right. Right was *yaus*, what one is bound to. A person accused of crime had to procure sureties, those who knew him, or members of his clan. As yet there were only freemen in the community; the dire curse of slavery had not arisen. Yet free laborers seem to have worked for hire. The community was on its road toward slavery. The system of human bondage has always made its appearance as an accompaniment of the growth of industry, the increase of fixed property, and the recognition of the value of labor as an element of wealth. Slaves would be useless to hunting tribes, and warlike hunters are apt to slaughter or burn their prisoners. To pastoral tribes they are of little more value. Their great use has always been to agriculturists. With the progress of agriculture prisoners speedily became too valuable to be slaughtered, and slavery steadily grew in its proportions, until in the great nations of Greece and Rome all the labor of the fields was performed by men of this class, and the noble art of war degenerated into a great slave-hunting raid. With the growth of commerce slavery has become again unprofitable, and a sentiment has been roused against it which promises soon to banish it from the earth. But the ancient community with whose history we are now concerned was as yet at the beginning of this great cycle which is now approaching its end. Only freemen existed in its midst.

We need not pursue this inquiry farther. We have sought to present a graphic picture of a vanished community whom we know mainly by our partial knowledge of the words it used. We have looked, through the lens of language, upon a primitive society, dwelling in barbarian rudeness and brutality, yet slowly advancing toward civilization, — a vigorous, energetic, strong-bodied, and active-minded race, stirring in body and soul, and destined to play a most important part upon the stage of the world. That we have given the whole story of their lives, cannot be affirmed. It was doubtless much richer than we can learn from our scanty stock of words. And much that we have said is open to doubt. Very likely many of the ancient Aryan words have died out of the languages of the modern nations and been replaced by other terms. Of those that have survived it is not always easy or possible to regain the original meaning, and it is quite probable that some of the interpretations adopted are incorrect. The ancient tribe lived a simple life, thought simple thoughts, and doubtless gave but a narrow and limited significance to its words. Yet that the picture we have presented is on the whole a faithful one there is little reason to doubt. And in the annals of mankind there is certainly nothing more remarkable than this rehabilitation of an antique community which had vanished ages before a thought of writing its history existed.

After the separation of the eastern and the western Aryans both branches advanced in knowledge and in the arts of life, and new words came into use. We may conclude with a brief glance at these new ideas and accomplishments as gained by the western branch. There arose among them extended ideas of family relationship. Words now came into use to designate the grandfather, the sister-in-law, and the sister's son. Terms of affection for old people arose. There was a similar advance in civil relations, and the lines of the community were drawn more closely. The citizen appeared as opposed to the stranger. A special act became necessary for members of one community to enter into friendly relations with those of another. In their industrial relations larger and better boats were produced. The sea acquired a name, and sea-animals, such as the lobster, the oyster, and the seal, became known. New plants and animals received names, — the elm, alder, hazel, fir, vine, willow, and nettle; the stag, lynx, hedgehog, and tortoise. Some of these were probably known before, but they had left no names. The duck seems to have now become domesticated; agriculture greatly improved. Millet, oats, and rye were cultivated. Peas, beans, and onions became common garden-plants. Terms for sowing, harrowing, and harvesting came into use. Yeast was used in bread-making. Glue and pitch became known; leather-work improved; the stock of tools increased; hammers, knives, shields, and spears were employed.

Yet with all these steps of progress the Aryans continued barbarians of no high grade. Manners were still rude, life coarse and hard, domestic relations harsh and oppressive, war bloody and brutal. The custom of tattooing and of painting their partly naked bodies with the blue dye of the woad-plant may have been common. They were yet rude barbarians, who had made scarce a step toward civilization. Such was probably the condition of the western Aryans when their later divisions took place and the existing peoples of Europe entered upon the historical path of their national development.

V.

THE HOUSEHOLD AND THE VILLAGE.

IT is our task now to review, so far as it can be traced, the general organization of the primitive Aryans, social, political, and religious. Our knowledge of the existence of this people has been gained mainly by the aid of language. But later research has opened several new lines of investigation, and taught us far more of the Aryan organization than that relating to its industries, habits, and possessions. Not only common words exist in all the branches of the Aryan race, but also common institutions, ideas, and beliefs; and by a co-ordination of these latter we are enabled to gaze deeply, through the shadows of time, into the very heart of that long-vanished community.

Not to go too far back into the origin of human institutions, modern research has made it plainly apparent that the germ of all existing social and political organization is the family. The domestic group appears everywhere as the seed of civilization, as it yet constitutes the unit mass of its organization. There is, it is true, another vital element in political development; but its influence has been of later date, and the family appears as the first clearly defined stage of condensation in the long upward progress of man from his very rude archaic condition. As to the gradual development of the family through its varied phases, embracing those of polygamy and polyandry, and monogamy with descent in the female line, to its final stage, with paternal headship and descent in the male line, the reader must be referred to works on that special subject such as those of L. H. Morgan and McLennan. It is sufficient for our present purpose to know that the Aryan family, at its earliest discoverable date, had attained the last-named stage of development, and as such formed the definitely constituted unit of the Aryan industrial and political organization.

Passing beyond the savage to the barbaric state of human development, we find the latter everywhere based on the family group. Alike in the agricultural tribes of ancient Asia and Europe, and in the hunting and agricultural tribes of America, this was the case. The monogamous family, composed of husband, wife, and their descendants, formed the unit of organization and the type of the clan and the tribal groups. In the pastoral tribes of Asia and the nations derived from them some degree of polygamy has always prevailed. Yet the first wife retains a position of special respect and authority, and monogamy is the rule with the great mass of the population.

In the early state of all the Aryan branches the family was organized under conditions of considerable similarity, — conditions doubtless inherited from ancient Arya. Each family, indeed, constituted a despotism on a small scale. The house-father was the head of the domestic group, and represented it in the community. Within the house precincts he possessed the governing power, and the right — if we may judge from the Roman example — to banish any member of his household, to sell his sons or daughters into slavery, to command them to marry whom he would, to seize on all their personal possessions, and to kill them at his will. It may be said, however, that some recent writers question the general absolutism of the Aryan house-father. It is certain, at all events, that his house was his castle. No one had the right to enter it without his permission, not even an officer of the law. It was his private kingdom, and for the acts of the members of the household he alone stood responsible to the community. The idea of personal individuality had not yet clearly arisen. The household was the primitive Aryan individual.

Such was the constitution of the family in ancient Rome, as declared in the extant Roman laws. The Roman father had the power of life or death over his children, and could banish them, sell them, or slay them at his will, and no man had the right to interfere. All the acquisitions of the son, all legacies left him, and the benefit from all contracts he made, were at the father's discretion; while he was bound to marry at his father's command. In the household the gradation of rank passed downward from father successively to mother, to sons, to daughters, to dependants, and to slaves; but the father was an absolute tyrant over all. In Greece the same conditions prevailed. K. O. Müller tells us that in Sparta the family formed an indivisible whole, under the control of one head, who was privileged from his birth. Cox, the historian, says that the house of each man was to him what the den is to the wild beast, into which no living thing may enter except at the risk of life, but which his mate and offspring are allowed to share.[1] In the Hindu family of to-day this inviolate character of the household is strictly maintained. A mystery overlies all its operations, — a remarkable se-

[1] Greece, p. 13.

crecy, which is maintained in the humblest households, and is probably a survival of a very ancient system of family isolation. With the Celts and the early Greeks there existed the right to expose or sell their children. This had become obsolete among the Teutons, though the right was recognized in case of necessity. With the Russians the power of the house-father, says Mr. Dixon, is without any check. He arranges the marriage of his son, makes the son's wife a servant, and stands above all law in his own house. His cabin is not only a castle but a church, and every act of his done within that cabin is supposed to be not only private but divine.[1]

Over one point alone the authority of the house-father was not absolute. He could do what he would with the movable property of the household and the labor of its inmates, but he could not sell or encumber the landed property. This was not individual, but corporate wealth. It belonged to the family as a whole, and was held inviolable. This was the law in all Aryan regions, from India to Ireland, with the possible exception of Rome, whose ancient laws relating to such matters are lost.

The heir to the family headship was usually the eldest son, though by no means always so. In Wales and in some other districts this office seems to have descended

[1] According to Wallace, this is rather the old theory than the modern practice. He remarks that "the relations between the head of the household and the other members depended on custom and personal character, and consequently varied greatly in different families. If the *Big One* was intelligent, of decided, energetic character, there was probably perfect discipline in the house. If not well fitted for his post, there might be endless quarrellings and bickerings." But there is every reason to believe that in earlier times the patriarchal power was absolute. "Russia," p. 88.

to the youngest son; and this is yet the rule among some of the southern Slavs. In default of a male heir one might be received by adoption.[1] The adopted son left his own household and became a full member of the new one, changing his tutelar spirits for those of his new family. The principle of adoption, indeed, was sometimes so extended in the clan as to make the claim of common descent extremely mythical. The whole Aryan system rested upon marriage and the birth of a male heir, who became eventually the head of the household, the system of family government being the type of the public organization. The ties of blood were scrupulously respected, and marriage among blood-relations forbidden to a greater extent than to-day. The wife became in every respect a member of the family group into which she entered, changed her household gods, and lost all obligations of duty to her former family, replacing them by new ties.

Such was the Aryan family, the antique political group from which outgrew the later clan organization of Arya. How it arose, with its peculiar feature of absolute domination of the head of the household, is not very clear. No such absolutism exists in the family group of the American Indians, which otherwise bears a very interesting resemblance to that of the Aryans, and Cox and Hearn trace it to a religious origin, — a duty resting upon the house-father, as representative of the departed ancestors, to pay due worship to their spirits and to manage the inheritance left him under responsibility only to these ancestral spirits.

[1] Under certain conditions the wife succeeded to the family government and care of the property, sometimes during the minority of the male children, sometimes during life if there were no direct male descendants. Maine's "Village Communities," p. 54.

This subject will be dealt with in the next section; and it will suffice to say here that the family group was apparently not limited to the living members, but included the dead ones as well, to whom sacrifice was offered, — perhaps as their share of the family food and wealth.[1]

In this religious duty we find a powerful check to the absolutism of the house-father. He represented the departed ancestors, and was answerable to them for a proper discharge of his duty. For any wrongful act he was liable to the vengeance of these powerful spirits, and might be exposed to dreadful calamities or become an accursed felon to the gods. It may here be said also that the power of public opinion was by no means absent from these ancient communities, and that it doubtless exercised a salutary influence over the acts of the domestic despot. The house-father was not expected to act by caprice, but to call a council of the family and of its near relatives to decide upon important matters; and very likely he was ordinarily governed by their decision. In this respect the family was the prototype of the clan.

Ancient as is the period to which we here allude, and vital as are the changes which have since taken place, the antique Aryan family, as a distinct political and industrial group, has not yet died out. It still exists in India and among the southern Slavonians, — the least progressed, politically, of the Aryan peoples. In India, in addition to the village communities, which form the ordinary industrial group, there exists a group known in Hindu law as the *Joint Undivided*

[1] The most dignified of the Indian courts has recently laid it down, after an elaborate examination of all the authorities, that "the right of inheritance, according to Hindoo law, is wholly regulated with reference to the spiritual benefits to be conferred on the deceased proprietor." — *Village Communities*, p. 53.

Family. In this the system of co-ownership is carried to its fullest extent. It is composed of the members of a single family, usually including several generations, by whom all things are held in common, — food, worship, and estate, — under the control of an elected head. This represents the primitive socialistic institution. The domain of the family is cultivated in common, the produce is held in common, and a common hearth and common meals are preserved through several generations. Significantly, in a region far to the west of this a closely similar institution survives. Among the southern Slavonians, in Croatia, Servia, and Dalmatia, the *House Community* is an ordinary institution. Here a single roof covers the family, which often comprises several generations and many individuals. The hearth and the meal are enjoyed in common, the lands cultivated by the common labor of the household, and all the produce held as the common wealth; the whole being controlled by an elected manager. These associations are not of recent formation and dissolvable at will, like their Hindu analogues, but have descended from far past time, each family continuing its organization, but sending out its surplus members, when they grow too numerous, to found other families. We can scarcely doubt that some of these Slavonian family groups have descended without a break from primitive Aryan times, and that they preserve to us, perhaps on original Aryan territory, the most antique form of the Aryan industrial group, which became replaced in later Arya by the institution of the village, next to be considered.

It may be here said that the limited duration of the Indian House Community — which rarely lasts beyond two generations — is due to the facility of dissolution under the modern Indian law. Originally it may have been as permanent as that of the Slavonic group. An interesting instance of a similar character, in a non-Aryan Indian tribe, is that of the Kandh hamlet, described by Dr. Hunter in his "Orissa." This people is still a nomadic one, and its institutions are strikingly like what those of the Aryans must have been in their specially pastoral age. The Kandh hamlet is a household unit in which individual rights are unknown. The house-father exercises supreme control, and the maxim is held that "a man's father is his god." Disobedience is the greatest of crimes. No son can possess property till the death of his father. Then a division is made of the land and stock, and each son becomes the head of a separate family.

The condition of society here reviewed is a highly archaic one, a survival from a very ancient period of Aryan existence when it was yet in the nomadic pastoral state. In its subsequent agricultural phase a different organization arose; but vestiges of the more ancient condition, in which the family was the state, persisted throughout this later period, and have, in the instances described, continued unto our own times. It is the patriarchal stage of political development, the stage which still persists generally among nomads, and which has played a remarkable part in the history of civilization, as we shall hereafter point out. The nomadic tribes of northern Asia and of the desert of Arabia are yet in this stage of organization. The principle of a single, supreme house-father has been there expanded into the head of the clan, the chief of the tribe, the ruler of the nation, through a direct process of development which has been modified by no secondary principle. The only Aryan people in which this archaic system has to any extent held its own in clan-government are the Highlanders of Scotland under their recent system of chieftainship. The Highland clan was a distinctively patriarchal organization, sustained by a people largely pastoral, and to some extent nomadic in habit. It was an expanded family group, in which the chief was the direct representative of the original ancestor, and was looked upon with a partly superstitious reverence by his ignorant and faithful followers. It seems to indicate a reversion to archaic political conditions.

In ancient Arya — probably when agriculture had begun to tie the former nomads to fixed locations, and to bring new interests into the foreground of men's thoughts — a new principle of organization gradually declared itself, a highly interesting outgrowth from the more ancient patriarchal system. This was the system of the Village Community, one of the most important stages in the development of human institutions. It must be borne in mind that with the acquirement of property in land industrial relations assumed a very different phase from that governing property in flocks and herds. In all these ancient cases the idea of community in property was firmly established. The common property of the family expanded into the common property of the clan, which was yet regarded as a single family, of common descent and common name. However greatly foreign elements came in, through adoption or otherwise, this fiction was maintained, and in several localities has not yet died out. There was no difficulty in sustaining this idea of community in the case of pastoral property. The herds were under the care of the whole group, and there was nothing to call for individualism in labor. And though they were held for the good of all, the patriarchal head of the group claimed certain supreme rights of ownership and management, and certain controlling powers over the clansmen, which were but a development of the original supremacy of the house-father. An interesting instance of such an organization is that of the patriarch Abraham and his followers and flocks as given in the Scriptures.

This generalism of duties could not so well be exercised in agricultural labor. Such labor could not properly be performed in common, and it became necessary to break up the tilled land into separate lots, each to be cultivated by a single family. This was attended or followed by the ownership of the product of its own lot by each family, although the land as a whole continued to be the property of the community. Instances of the growth of this system may be found in American institutions. In the Inca empire of Peru the system of agriculture and government continued patriarchial in great part. The population as a whole cultivated the lands of the Inca and the Church; the products, though held in part for the good of the people, being under the supreme control of the ruler. But the remainder of the lands, those specially appertaining to the people, were divided into separate lots, each cultivated for its own use by a single family. In the Aztec empire of Mexico the supremacy of the Montezumas was much less absolute. The lands were partly claimed by the Throne and the Church; but the work on these lands was done by dependants, not by the people as a whole. The remaining lands belonged to the separate cities or districts, and were divided among the people. But a part of all produce went into the public storehouses, and was under the control of the government. Among the partly civilized tribes of the southern United States — the Creek confederacy and the adjoining tribes — all the land was the property of the people, and was divided into separate lots, apportioned to the separate families, though some degree of individual ownership was also exercised. But a portion of all produce, alike of agriculture and of hunting, was obliged to be placed in certain public storehouses for the use of the people in case of necessity. These public stores were under the supreme control of the *mico*, or village headman, in whom we have a close representative of the similar officer in Aryan communities, though the *mico* had besides an important spiritual authority.

Coming now to the Aryan organization, we discover the final stage in this gradual separation of interests. Here also the land as a whole is the property of the community; but it is divided among the families for their separate use, and all trace of community in its produce is lost. The wise system of public storehouses of the Indian village does not exist, and the product of each separate field is the sole property of the family cultivating it, to be disposed of without supervisal. Thus in these several peoples every stage of growth, from the pastoral complete community in cattle to the Aryan partial community in land, can be traced.

It is to this separation of interests in the common property that we must look for the origin of that peculiar clan-organization which is, in nearly a complete sense, a special characteristic of the Aryan people. In this organization the individuality of the family persisted. There was no merging of the smaller into a larger patriarchal family group. Each household became an equal unit of the village group, with equal rights in the common property, and with an equal voice in the decision of all questions relating to the general interests. The head of each family was a full member of the community, and the government was in the hands of these freemen, organized into a council. So far as we can discern, this was the archaic condition of the village community. The tendency to continue the patriarchal organization had been checked by the division of interests, and the separate yet equal rights of every freeman in the common property. The principal questions necessary to decide related to industrial affairs, and in the disposal of these every house-father had acquired an equal right.

Yet the patriarchal tendency was checked, not killed. Old ideas have a persistent vitality in barbarian communities. The members of each village viewed themselves as kindred, descendants of a common ancestor, and in each village there were certain families which were regarded as more directly in the line of descent from the ancient ancestor. A certain gradation of rank existed, dependent on honor, not on privilege; and when it became necessary to choose a leader in war, or to elect some umpire in village disputes, the choice most naturally fell on those deemed to have a hereditary claim to authority. The offices of chieftain and of village head-man thus arose. The village was constituted on the type of the family. In the latter a council was called to decide important affairs, and in certain cases to elect a family head. It was the same with the village. The council of freemen held the rights of decision and of election; but in both family and village the choice usually fell on those having the best claim of hereditary right, and the election often became a mere acclamation in favor of the person recognized as the natural chieftain.

All this is not mere conjecture. There is abundant historical evidence of the organization of the ancient Aryans. It was evidently at once a communistic and a highly democratic society. In its latter characteristic it was markedly different from the patriarchal society, which was aristocratic in tendency, and which naturally tended to despotism; while in all Aryan communities the ancient claim of equality of rights and privileges has had persistent vitality, even under grinding despotisms. All modern democratic governments are direct outgrowths of the ancient organization of the Aryan village, while the despotisms of Asia are as direct resultants of the patriarchal system.

One statement more is necessary in regard to the division of property in ancient Arya ere we adduce the historical testimony. Each village claimed the right of eminent domain over a landed district of definite extent. But in the management of this landed property there were three separate interests to be considered, — the pastoral, the agricultural, and the domestic. It is interesting to observe the disposition of these. The pastoral interests retained their old generalism. The pasture-lands were held in common, for the feeding of the flocks of the villagers. The arable lands, on the contrary, were equally divided among the several families for cultivation. But, as if to prevent any claim to individual ownership, these lands were periodically redistributed. This system of redistribution is still maintained in Russia. Finally, the village plot was divided into house-lots, which were the absolute domains of their proprietors. Each family held separate ownership in its house and the plot of ground surrounding, and perhaps partly for that reason jealously guarded it. Each man's house was his stronghold; it was the only spot of earth in which he could claim individual ownership; and every man who attempted to intrude on it without his permission was an enemy whom he might repel as he would his deadliest foe. Possibly this may have had something to do with the growth of that isolation of the household which became so strongly developed in all Aryan communities.

If now we come to look for the historical evidences of this assumed industrial and social status of the ancient Aryans, it is remarkable, considering the numerous and radical changes in human institutions since the opening of the historic period, what clear traces of it remain. We have already described the extant relics of a yet older Aryan condition, — that of the patriarchal family. The clan system has been equally persistent, and exists with little change in Russia and India to-day, while historic traces of it can be found in every other Aryan community, with the exception of that of Persia; and even in Persia the ancient democratic organization of the people can be clearly traced.

There is considerable evidence that the ancient Hellenes and Romans were organized in village clans, with common landed property. Morgan says that the Athenian *gens*, or clan, in some cases, at least, held property in common. Thucydides speaks of such communities as independent systems of local government, and there was seemingly a period in which there was no city of Athens, but many village communities in Attica. The Roman gens was similarly in possession of common lands, of a common clan-name, and of common religious rites, burial-place, etc. Mommsen describes " village communities by the Tiber," out of which Rome arose. There is no doubt of the existence of such clan villages. The hills of Rome and the

Acropolis of Athens formed originally centres of refuge for the villagers in periods of invasion, and it is supposed that in such hill forts we have the germ of many of the ancient cities. The modern city of Calcutta had its origin in an aggregation of several separate village communities.

The Celtic Aryans present similar indications. The sense of kinship is deeply stamped on the Brehon laws of ancient Ireland, and the Irish *sept* probably repeated the joint family or the village clan of the Hindus. Private ownership in land was common at the earliest historic period, yet the rights of private owners were limited by the communal rights of a brotherhood of kinsmen. Apparently the original right to cultivate a fixed plot was then growing into a claim of private ownership in that plot, as became the case elsewhere. The power of the lord of the manor over the communal lands was also beginning to show itself. The *fine* or *sept* bore the name of its supposed ancestor, and its territory also bore his name, — a condition which has not yet died out. As elsewhere, the sept received strangers by adoption; but this did not destroy the fiction of kinship.

In Scotland the village community was a much more persistent institution. It left its marks as late as the time of Sir Walter Scott, who discovered traces of such an institution in the islands of Orkney and Shetland. Very recently, in the Lowlands of Scotland, in the borough of Lauder, a condition of affairs has been discovered closely analogous to the antique village community system.[1] Sir Henry Maine has also traced in France an indication of a like condition of affairs, despite the violent revolutions to which that country has been subjected.

[1] Maine's Village Communities, p. 95.

The facts relating to the Teutonic village communities, as traced by Von Maurer in his valuable series of works on the subject, and of vestiges of the same institution in England, as shown by Nasse, may be here epitomized. The ancient Teutonic agricultural group consisted of a number of families holding a certain well-defined tract of land. This tract was divided into three portions, known as the mark of the township or village, the common mark, or waste land, and the arable mark, or cultivated area. These three sections were held under very different conditions. The waste was the common property of the community, held for purposes of pasturage, for gathering fire-wood, and the like. It was the analogue of the old pastoral domain.[1] The village section was divided into house and garden plots, each the sole property of the family occupying it. No one, not even the officers of the law, had the right to intrude upon the family domain. There the house-father was absolute lord. The arable mark seems in almost every case to have been divided into three great fields, only two of which were cultivated in any one year, the third lying fallow. But tillage was not in common. Each house-

[1] The waste formed the line of demarcation between different communities, — the wooded region of the hunter, the hostile border-land which the foot of the invader must traverse. We have survivals of the word which designated it in *Denmark*, or the Danes' Mark ; in the *March* or battle-border between England and Wales ; and in the *marquis* or *markgraf*, the guardian of the mark. The waste mark was also the seat of exchange of products between villages, the region of the *market*. The forest of the waste was the temple of the Teutons, the home of the unknown and uncanny, of ghost and goblin. It was the least-known and most-dreaded of their dominions. Here dwelt Odin, the god of the mark, the spirit of the tree and the forest breath, the god of the wind and the tempest. Within the village domain dwelt order and peace ; there man was master. But in the waste land beyond, terror was lord, and the supernatural held high carnival.

holder had his family lot in each of the three fields, which he tilled by his own labor and that of the members of his family, while he had absolute rights in the disposal of its produce. But he could not cultivate as he pleased. He must sow the same crop as the rest of the community, and observe fixed rules as to modes and times of cultivation. Nor could he interfere with the rights of other families to sheep and cattle pasturage in the fallow lands, or in the cultivated lands after the harvest. The rules of custom governing the common interests were very intricate, and extended to minute details. Many of them had come down from very ancient times, while others were formed as new questions arose. There was little difficulty in enforcing them; they had almost the force of sacred laws. The main evidence of gradual change we can discover is that from the antique periodical redistribution of family lots to the continued cultivation of a single lot, and finally to the restrictive ownership of this lot.

As to ancient evidences of this condition, we may quote from Cæsar, in his description of the Suevi (Swabians): "They have no private and separate fields," and "none have fixed fields and private boundaries, but the magistrates and princes in assembly annually divide the ground in proportion and in place among the people, changing the arable land every year."[1] Tacitus gives testimony to the same effect, saying that the lands were held by the farmers in common, and the fields occupied in rotation. "They change their tillage land annually, and let much lie fallow. . . . They do not hedge their meadows, nor water their gardens, and they cultivate only corn."[2]

[1] De Bello Gallico, iv. 1, and vi. 22.
[2] Germania, 25–26.

It is a striking evidence of the conservative persistency of institutions among agriculturists to find that similar conditions exist to-day in middle and south Germany, with but slight modifications. The main change is that communism in the arable lands has ceased, and the fields of the peasants are held in private ownership. The valuable work on Germany by Baring-Gould gives some interesting information and suggestions on this point. He makes it clearly evident that the customs of the Aryans changed in accordance with the variation in the character of their soil. Where the land was poor, as in northern Germany, it was incapable of supporting a dense population, and such regions became active centres of migration. The seeming general migrations were in reality only partial, and mainly consisted of the swarms of elder sons whom the paternal estates could not support. In such cases but one son remained under the paternal roof, perhaps in some cases the eldest, but oftener the youngest, — from which may have arisen the custom in some localities of inheritance by the youngest, as already mentioned. Such was probably the origin of the frequent invading movements of the Saxons, Angles, Franks, etc. Room for the surplus population was needed, and they obtained it by conquering a new home, or died by the swords of the invaded people. It was a system of the survival of the strongest which served to settle the Malthusian difficulty during long ages of human history.

In southern and middle Germany, where the land is richer, the communal conditions more fully prevailed. In the North the farm developed, descending to one son as the heir, — a condition which still prevails in that locality. In the South the village persisted, with its common lands.

This system was nearly universal among the Franks, Alemanni, and Swabians, and survives unchanged in some places. Thus at Gersbach, in the Baden Schwarzwald, all the tillage land is held in common and is periodically redistributed. In the Altmark all the land is common, and the agricultural work to be done the next day is decided every evening by the heads of households. Similar conditions exist in other places. The three-field system is yet universal in this region, and in numerous cases the pasture and forest land is still held in common. The *Gewannen*, the village arable fields, consist of somewhat narrow strips, divided from each other by footpaths. These are subdivided into still narrower family strips, marked off by trenches or stones. They are usually rectangular, often not more than seven yards wide, and in extreme cases reduced to three or even one yard in width. In such cases they are longer in proportion to their narrowness. These fields are divided into the *Feld*, the *Flur*, and the *Zelg*, the winter, summer, and fallow field, in accordance with immemorial custom. The lots of peasant proprietors are thus divided into narrow strips scattered all over the parish, such a thing as a compact farm being very rare. Of recent years, however, efforts have been made by the Governments to end this state of affairs and redistribute the land so as to bring each peasant's holdings together. The indications are that ere long the old and inconvenient system will vanish under the force of modern ideas and governmental initiative.

That the soil of England was originally divided in a similar manner by its Saxon conquerors we have abundant evidence in the many traces of communistic agriculture which still exist. Fields known as "common fields" may yet be found in many of the English counties. These fields are nearly always divided into three long strips like the German *Gewannen*, separated by green baulks of turf. The separate farms consist of subdivisions of these strips, often very minute. There is evidence to show that the same owner once held a share in each strip, and that these shares were equal, or nearly so, though now many of them may be accumulated in single hands. The methods of agriculture closely reproduce those of old. One strip is left fallow, while unlike crops are cultivated in the other two strips. The right of common pasturage for the cattle of the farmers often exists; and the shares in the arable lands in rare cases shift owners annually, as in old Arya. This is frequently the rule with the meadows, rights in which are often redistributed annually by casting lots.[1]

In addition to these arable fields there are in many parts of England open or common fields, sometimes comprising more than half the area of certain counties. Mr. William Marshall, in his "Treatise on Landed Property," estimates that a few centuries ago nearly the whole of the lands of England lay in this open state, and formed the common property of cultivators. They seem to have been divided into arable and waste or pasture lands on a principle closely related to that of the Teutonic village. Similar conditions yet exist in Lowland Scotland, as in the borough of Lauder, already cited.

This persistence of the communistic village organization in England, after all the wars and revolutions in that land, shows a peculiar vitality in the ancient Aryan system of property holding. Significantly similar institutions were established in America, the yeoman settlers of New Eng-

[1] Maine, Village Communities, pp. 78 to 89.

land dividing their new soil on the principle to which they had been accustomed at home. These American village communities, however, never took a deep hold on the soil. The flood of new emigrants soon drowned them out of existence.

In two Aryan lands, India and Russia, the village community has been rigidly persistent, and exists at the present day in a form not widely different from that which must have prevailed in ancient Arya. Only among the Hindus and the Slavonians does the archaic house community persist, while they everywhere maintain the village system. The Indian village closely repeats the Teutonic, as above described. There is the arable domain, divided among the families, yet cultivated under minute laws of custom. Where grass-crops can be raised, the meadows persist, on the verge of the cultivated ground. Outside appears the waste, the undivided pasture-ground of the villagers. Centrally lies the village, with its individual family plots and its strictly isolated households. And all is under the control of an elected headman or a village council which decides all questions. Two ancient ideas have died out, however. The periodical redistribution has disappeared, except as a tradition, and the villagers do not consider themselves kinsmen. Perhaps the abundant infusion of foreign blood has killed out this old conception.

The old system of government by an assembly of adult males, as found in the ancient Teutonic community, has partly vanished in India. In many cases the affairs of the community are managed by a council of village elders, but more generally this council is replaced by a headman, — a feature of later origin. This office is sometimes hereditary, sometimes elective; though in the latter case usually confined to a particular family, and generally to the eldest male of that family.

The Indian villages are not solely cultivating communities. Manufacturing interests are also included. There are families of hereditary artisans, as the blacksmith, the shoemaker, etc. There is a village accountant, a village police, and other necessary officers. But these persons are included in the communistic system, and are paid by an allowance of grain or a piece of cultivated land. All their wares have a price, fixed by usage, and to bargain with a Hindu tradesman for his goods is to insult him.

In central and southern India are certain villages to which is attached a class of persons who form no actual part of the community. These persons are looked upon as impure. Their touch is contaminating. They are not permitted to enter the village, or only a reserved part of it. Yet they have definite duties, one of which is the settlement of boundaries. They probably are descendants of the aboriginal population. Still, despite the rigid exclusion of these "outsiders," there can be no question that the alien population largely made its way into the village in past times, as is shown by the evident great mixture of race-characters in India, and by the loss of the idea of kindred in the village groups. In the Russian community this is avoided by the ease of swarming to new lands. But in densely peopled India the contest between the group of kindred and the alien class for a share in the land must have been severe and persistent, and to it is probably due the conditions we now find.

Of all modern Aryan nations, however, Russia is the one that has deviated least from the ancient customs, and in the Russian *mir* we have the closest analogue of the antique Aryan village. This is in accordance with the view we have taken of Russia as the Aryan branch that has remained nearest to or yet occupies the primitive home of the race, and that has been least exposed to disturbing influences. Yet the unwarlike character of the Russian, as of the Hindu peasantry, and their close confinement to agricultural duties, have doubtless had much to do with their strict conservatism. In all lands and in all times the agriculturist has been the conservative, the citizen the radical; while but for the disturbing and destroying influences of war we might have to-day the most archaic of institutions persisting in their full vigor.

In Wallace's admirable work on Russia is an interesting description of the Russian *mir*, or village community, which may be here epitomized. Ivanofka, a village in northern Russia, is offered as a typical instance of a cultivating group. It embraces in its communal bounds about two thousand acres of a light sandy soil. In the cultivation of this nearly all the women and about half the males of the village are habitually engaged. The land is separated into three portions, — arable, waste, and village; the arable being divided into three large fields, after the immemorial Aryan usage. The first field is reserved for the crop of rye; the second for oats and buckwheat; while the third lies fallow, and is used as pasture-ground. This distribution changes from field to field annually, so as to make a rude rotation of crops and to give each field rest one year in three. The fields are cut into long, narrow strips, of which each family possesses, according to its needs, one or more in each lot. Many of the villagers are artisans, and live in the towns. Yet they cannot leave the village without consent of the council, must return to it when ordered, and must send part of their earnings home to the village treasury. Otherwise they forfeit their hereditary claims, and break a link of connection with the ancestral home and kindred which is dear to the heart of every true Russian.

The chief person in the *mir* is the *selski starosta*, or village elder, whose office is elective, and presents no trace of heredity. The electing body is the *selski skhod*, or village assembly, composed of the adult members of the community. This body settles all important affairs. As the power of the elder here is limited, so is that of the house-father. He has in recent times lost much of his ancient absolutism, and no longer rules with unquestioned authority over the adult members of the family. The affairs of the village are closely regulated by custom. No one can plough or mow until the assembly has met and passed a resolution, and no peasant dreams of disputing a decree of the assembly. These decrees are generally carried by acclamation, though there is a counting of heads by the elder when any diversity of opinion appears. And it may be said that no one desires the office of elder. It brings with it trouble and responsibility, with very little compensation. Efforts are made to avoid the empty honor, though no one dare dispute the decision of the electors.

In regard to the division of the fields among the householders, the principle of periodical redistribution is yet extant, and is practised whenever changes in the number and size of families make it desirable. And the idea of kinship still persists. The Russian villager believes himself allied by blood-ties with the members of his village group. In the more fertile southern districts each peasant strives to obtain all the land he can get, — which is not the case in the North, where the land-tax renders too large a farm undesirable. All disputes thence arising are settled by casting lots. In these districts the meadow-lands are also divided into household shares; but this division is made annually instead of irregularly, as in the case of arable lands. Occasionally the grass is cut in common, and then divided. It may be said, in conclusion, that the meetings of the assembly of the village are very informal, and discussion is carried on in a free and easy way, though with considerable shrewdness. Wallace gives some very amusing instances of these debates, — the direct counterparts, probably, of the methods of government that prevailed in ancient Arya centuries before history was born.

The village community, however, while found universally among the Aryans, cannot be claimed as a peculiar Aryan institution. It is one of the two forms under which all ancient agricultural societies seem to have been organized; the other being the more archaic patriarchal system. Village communities have been discovered in Java and among North African Semitic tribes, while they form the ordinary type of the Indian clan groups of North America. It has been the custom to speak of the Indian tribes as in the hunting-stage of development. But the fact is that they were very largely agricultural. For one evidence of this the reader may be referred to a paper in the "American Naturalist" of March, 1885. And their land-holding customs, together with their system of organization, bore a striking resemblance to those of the Aryans, though with some features of variance, as will be seen when we come to treat of their comparative political systems. This much may be here said, — the idea of kinship in the clan was strongly held by the Indian tribes, but the isolation and rigid exclusiveness of the household was not maintained. The belief that "every man's house is his castle," to be defended to the death if need be, is peculiarly Aryan. Its counterpart is found nowhere else in the world.

VI.

THE DOUBLE SYSTEM OF ARYAN WORSHIP.

IN the religion of the ancient Aryans is displayed, to a more marked extent than in that of any other people, two distinct systems of worship, arising from unlike influences, and struggling for precedence. This fact is of importance, as it has had a vital influence on the history of their descendants, and has done much for the preservation of their democratic spirit. For of these two systems the one tended to aristocracy, the other to democracy; and in nearly all the ancient Aryan communities the democratic religious system kept the ascendency.

We are apt, indeed, in considering the Aryan religions, to call up before our mental vision simply the rich picture of mythology, with its intricate and extraordinary details, its surprising variety of conceptions, the physical splendor of its deities and their habitation, and the crowding multitude in which they inhabited earth, air, ocean, and the over-arching skies. But these marvellous mythical deities were not the oldest or the most venerated gods of the Aryans. They grew into great prominence in the early literary period of Greece and India and of the Teutonic tribes, and became surrounded with a confusedly complex series of biographical details, in which the vestiges of their origin were lost to their worshippers. But in ancient Arya the nature gods lacked this complexity of myth and variety of forms and attributes, and their true meanings were plainly apparent. They were as yet the sky, the sun, and the planets, the winds and the clouds, the summer and the winter, the dawn and the darkness, and those varied elemental phenomena which are of supernatural significance to the simple fancies of all uncultured peoples. They had not yet unfolded into the Supreme Deity of heaven and earth, with his brilliant and marvellous court of secondary immortals.

Less striking, yet more ancient and more persistent, than this system of worship was another, of which we see and hear but little, yet which formed the most generally observed religion of our far-off progenitors, so far as indications prove. This was the worship of ancestors, the home-worship of the Aryan family, the exclusive worship of the Aryan clan, the religion of the hearth and of the ancestral tomb, — the only worship that really reached the hearts of the early Aryans.

Something very similar to the Aryan religious system exists to-day in China as a phenomenon that has utterly died out elsewhere in civilized lands. There, too, we find a double system, — the worship of ancestors underlying the more public systems of belief. But the Confucian philosophy has never taken deep root as a popular religion, while ancestral worship has a stronger hold on the public heart than Taoism or Buddhism. On the Western continent, among the Indian tribes of the southern United States, appears a similar double system. Here, however, it was not an ancestral, but a demonic system, a developed Shamanism, that was mingled with the worship of the elemental gods. But while the worship of ancestors held the supremacy in China, that of the solar deity and of later mythical gods did so in America. Among the Aryans it is probable that there was a closer balance of influence between the two systems of worship. Very probably in ancient Arya ancestral worship was strongly in the ascendant. Later it became to some extent balanced by the growing prominence of mythological worship. But the latter attained supremacy only in India and perhaps among the Celts. Elsewhere the indications seem to show that the former continued the dominant system.

In considering this question we are dealing with one of which the history is somewhat obscure. The Aryan house and clan worship did not attract the attention of the poets, whose verses are filled with the marvels of mythical legend. The family worship was in no sense public, like that of the elemental deities. It was conducted in secrecy and mystery. Strangers were not admitted to the sacred rites of house and clan. And every family had its own ritual, which was a secret never to be divulged. In consequence very little testimony concerning this system of worship has made its way into literature. It is only alluded to incidentally, in vagrant paragraphs; and what little is known of it has been recovered only by patient research and by piecing together flitting fragments of evidence. Necessarily, to some extent, doubt creeps in. We can rebuild the ancestral worship only in outline. It has nowhere in the past been made the subject of brilliant essays and the groundwork of great poems, like those devoted to the multitudinous deities of mythology.

The worship of ancestors seems to have been almost universal among mankind in a certain stage of development. Traces of it can yet be found in all parts of the earth. But, so far as appears, it became a well-defined and largely exclusive system only among the Chinese and the ancient Aryans. And it is in all probability to this worship of its ancestors by the members of the Aryan household that we owe the peculiar secrecy of family life, the supremacy of the house-father, and the strong resistance to intrusion upon the domestic domain. According to the theory of Cox, the original ancestor of the family became a deity whom the survivors had to worship and propitiate. His burial obsequies needed to be duly performed, and rites of sacrifice to be paid to him. This could be done only by the eldest son, his legal representative. Thus the house-father became the house-priest, and the continuance of the family a religious necessity. To let it die out from lack of offspring would have been impious, and to this was due the practice of adoption, in default of male heirs, which afterwards became so extended a custom in the Aryan clans. But the tendency was to reduce every kind of association to that of kinship; and this idea was kept up long after the free adoption of strangers had rendered it an utter myth. To the position of the father as the family priest and the offerer of rites to the ancestral deity, whom he represented, we owe his supremacy as the family ruler. The family was a composite one, made up of several generations of the living and the dead, of all of whom the house-father stood as the central point. It was a sacred group, which it was his duty to keep together, and to suppress all insubordination that might threaten its integrity. Doubtless from the position he thus held gradually rose his absolute power and the unquestioning submission to his decrees. He spoke with the voice of the whole body of ancestral deities, and was responsible to the house-gods for the rightful performance of his sacred function.

Hearn, in his "Aryan Household," has given a highly interesting description of this ancient system, which we may here epitomize, at least in its more trustworthy details. Kinship and community of worship and property were the ties which first bound men into definite groups, the family bond expanding into the first national bond, — that of industrial and religious communism. It began with the family, extended to the clan, and thence to the tribe, attaining a very considerable extension before it was replaced by the territorial system of civilized nations. Each family had its common burial-place. This in later times became the common burial-place of the clan or gens, in which it would have been sacrilege to inter a stranger. In very early times it is probable that the bodies of deceased ancestors were interred in the dwelling. At a later date they were kept for some time in the dwelling, and then interred outside. These customs are still in vogue in China. They gave the deceased a very close relation to the house, and to a very late period the hearthstone seemed to be considered in the light of an altar to the ancestors, the sacred stone of oblation to the departed.

The common meal was apparently the symbol of the common worship, though probably this symbolic significance was only recognized in meals specially prepared in honor of the dead. Spirits could not be expected to come unless specially invited and their share set apart. Yet they did not consume the gross part of the food, but only its spiritual essence, — all objects being supposed to have souls. In this we seem to have the origin of sacrifice, while the after-consumption of the food by the priests was but a sharing in the holy banquet, of which the deities had regaled themselves on the spiritual portion. Many illustrations might be drawn from ancient history of such sacred feasts to the deities of families and clans, and feasts to the dead are celebrated in Russia to the present day.

The evidences of this ancestral worship are abundant. The Hindu Vedas distinctly recognize the worship of the *Pitris*, or fathers, and to this worship the Sama-Veda is specially devoted. "The *Pitris* are invoked almost like gods; oblations are offered to them, and they are believed to enjoy in company with the gods a life of never-ending felicity."[1] A similar belief existed among the Iranians, who worshipped the *Fravashis*, or spirits of the dead, and especially of their own ancestors. The latter worship was conducted with strict privacy. With the Hellenes the family worship of the house-spirits — the "Gods of the Hearth," or "Gods of the Fathers" — was common. On the Romans it had a specially deep hold, and reduced the public worship almost to a nonentity. For these house-spirits we have many names, — the Genius, Lares, Penates, Manes, and Vesta. Vesta was the hearth, with its holy flame. The Lares and Penates were the true house-spirits, the ancestral gods so dear to the Roman heart. We know little about this family worship with the Slavs,[2] Teutons, and Celts. We have no ancient literature from the pre-Christian days of these peoples. Strong efforts were made by the Christian Church to abolish every phase of heathen worship, yet it has not succeeded in suppress-

[1] Max Müller, Chips from a German Workshop, ii. 46.

[2] Ralston tells us that "the worship of the Slavonic Lares and Penates, who were, as in other lands, intimately connected with the fire-burning on the domestic hearth, retained a strong hold on the affections of the people even after Christianity had driven out the great gods of old." — *Songs of the Russian People*, p. 84.

ing all traces of the ancestral deity, — which indeed has left its mark in the guardian or patron saint of the Catholic devotee, and in the feasts to the dead among the Slavs and elsewhere. With the Russians the ancient family god yet lingers as the *Domovoy*, — the house-spirit, or angel in the house; reproducing the "hero in the house" of the Greeks, the Roman "man in the house," and the Teutonic *Husing*. Among the Teutonic nations, indeed, there are many traces of the house-spirit in its later form of a half-demonic goblin. We have it in the *Hausgeist*, the *Kobold*, the Brownie, the Robin Goodfellow, etc., — prankish elves, ready to do the house and hearth work of neat housekeepers during the night, but apt to leave annoyance for the idle and careless. These house-goblins could be propitiated by offerings left them, — probably a relic of the ancient sacrifice. But they became the foes of those who neglected them, as the ancient house-spirits became the deadly enemies of those who failed to offer them due libations. In short, as to the general existence of ancestral worship, either as a persistent fact or as a transformed survival, we may quote from Tylor: "In our time the dead still receive worship from far the larger half of mankind." [1]

The Aryan house-worship seems to have been conducted with inviolable secrecy. Each family had its own ritual, which was a precious secret, never to be divulged, and which appears indeed to have had the force of an amulet. Thus in the Rig-Veda the antique poet sings: "I am strong against my foes by reason of the hymns that I hold from my family and that my father has transmitted to me." In Greek legend we find that Polyphemus scorns the authority of Zeus; he will recognize no god but his

[1] Primitive Culture, ii. 112.

own father, Poseidon. So the Russian peasant of to-day draws a line of distinction between his own *Domovoy* and that of his neighbor. The former will aid, but the latter will seek to injure, him. The ancient house-spirit was the house-guardian, who repelled thieves and warned trespassers. Little the ancient Aryan cared if the universe had one or many authors. The gods of his own hearth were nearer and dearer to him than these remote deities of all mankind.

As the Aryan family expanded into the Aryan clan, so did the house-worship into that of the clan, whose rites were paid to the remote ancestor of the group of kindred. It is a question of some interest to what limit of ancestry the family worship extended. Mr. Hearn thinks it was limited to the great-grandfather, and that the household might be made up of six generations, three of the living, and three of the dead. At this point, in his view, the house unfolded into the clan, colonists being sent out to found new households, and the immediate kinship of the family being exchanged for the more remote kinship of the clan, while the common deity worshipped by the several families was the spirit of the ancestral founder of the clan. It is doubtful, however, if any such definite rule prevailed; and no doubt inclination or internal disorganization had much to do with the disintegration of families and the growth of the wider and less intimate association of the village or clan. The existing Chinese custom is of interest in this connection. As a rule the Chinese family worships the spirit of the father and the grandfather. But this home-worship never seems to extend beyond the third generation of the dead. The Chinese clan, on the contrary, worships its remote ancestor whenever known, and the grave of such an ancestor, if preserved, forms a sacred centre for the religious services of the clan. The descendants of Confucius, for instance, worship their great ancestor to-day as the chief of the gods to them.

So the Aryan clan-worship was as devoted and as exclusive as that of the family. Special gods of tribes and clans existed among the Teutonic and Celtic tribes, while the worship of the ancestor of the gens was a common custom with the Greeks and Romans. Mr. Hunter tells us that it is the first duty of a good Hindu to worship his village god.[1] Among the Semitic tribes evidences of the same custom exist. The Bible, in its story of the Hebrew patriarchs, yields testimony to this effect. With the Aryan clans this worship was secret and exclusive. A strong feeling existed against intrusion on the sacred rites of a Greek or Roman gens. We are told, indeed, that the presence of a stranger at the religious ceremonies of a Greek clan was intolerable. And these ceremonies seem to have been held at the common burial-place of the clan, — a strong indication that the worship was paid to the original ancestor. All these ceremonies, however, were conducted with such secrecy that we know very little concerning them. There seems to have been a dread that a god might be stolen or seduced away if not guarded with strict care. For this reason, perhaps, the name of the tutelary deity of Rome was always kept a profound State secret.

On the other hand, the worshippers might reject or desert their god, if found weak to redress their wrongs or to protect them from evil. Several amusing illustrations of this may be given. The Finns of to-day in time of need do not hesitate to neglect their gods and pray to the more

[1] Orissa, i. 95.

powerful Russian deities. So we are told, as an incident in Roman history, that "the statue of the Cumæan Apollo came near to being thrown into the sea, from an ill-timed fit of weeping. Fortunately it was considered that the tears were for his old friends the Greeks, not for his new friends the Romans."[1] As a more modern instance we may quote: "A prince of Nepaul, in his rage at the death of a favorite wife, turned his artillery upon the temples of his gods, and after six hours' heavy cannonading effectually destroyed them."[2]

It was this secret, domestic, and clannish worship of the Aryans that hindered the public worship from gaining a controlling influence, and checked the growth of a powerful priesthood in most branches of the race. There was not the almost complete hindrance to the growth of mythology that we find in the early Chinese; yet the worship of ancestors was sufficiently strong to prevent mythology from becoming dominant as a religion. Beneath it, almost unseen by us, yet vital and vigorous, lay the more ancient system, that of the worship of family and gentile ancestral gods. Yet ancient Arya was not without its other deities. Its people possessed an active imagination, and could not avoid being vividly impressed with the mighty powers and strange phenomena of Nature, which they naturally endeavored to explain or comprehend. And, as in every ancient effort at such explanation, they arrived at the conception that these phenomena were the work of intelligent and powerful beings, the overruling gods of earth and heaven. In the primitive era they had nothing that can fairly be called a mythology. They worshipped Nature as

[1] Saint Augustine, City of God, i. 101.

[2] W. E. Hearn, The Aryan Household, p. 25.

they saw it, with no idea of symbolism and no misconception of the meaning of their objects of reverence. It was yet summer and winter, daylight and darkness, the bright dawn and the terrible storm, thunder and sunshine, which they looked upon as the powerful deities of the universe, and upon whom they called for protection, or whose dark wrath they deprecated in cases of peril beyond the power of their humbler domestic deities. Only by slow degrees did these elemental gods lose their original significance. Probably at an early period the Aryan imagination had begun to invest them with metaphorical significance. The Clouds became the cows of the gods, whose milk refreshes the earth, but which at times are hidden in caves by robbers. The Dawn, the beautiful spirit, sends her glad eye-beams over the earth, and is speedily pursued by the glowing Sun. In winter the Earth mourns for the dead Summer, which lies buried in the dark prison of Hades. Or the Summer sleeps in the land of the Niflungs, the cold mists, guarded by the serpent Fafnir, while her buried treasures are watched by the dwarf Andvari. Hundreds of such metaphors gradually grew around the movements of the sun, the winds, and the clouds, the demon Night, and the bright god Day, the all-destroying Winter and the all-restoring Summer. In time the origin of these metaphors became obscured, and even the derivation of the names of many of the gods was forgotten. Mythology gradually rose out of the primitive worship of the powers of Nature, and the endless biographical details which surrounded the mythologic deities testify to the original activity of the Aryan imagination.

An interesting feature in the primitive Aryan mythology is the selection of the bright, broad arch of the heavens as the primal deity, the great father-spirit of gods and men. This deification of the sky was not peculiar to the Aryans. We find traces of it in Babylonian, Chinese, and American worship. But at a very remote period in the civilizations of Egypt and Babylonia, Mexico and Peru, the sun gained supremacy as the first and greatest of the gods, the prime spirit of the universe. With the Aryans the sun was much later in attaining acknowledgment, and the shining arch of the sky continued the deity supreme. This is the deity that descended to historic times as the great father-god, the object of highest reverence to most of the Aryan peoples when first they emerged into history. Varuna, the elder god of the Vedas, was the veiling heavens. He stands opposed to his brother Mitra, who is the deity of the noontide sky, while Varuna appears to represent the starlit firmament. We find this god again in the Uranos of Greek mythology. He sits, in the words of the Vedic poet, throned in splendor, clad in armor of gold, and in a palace supported on a thousand columns, while around him stand ready the swift messengers of his will. At a later date another heaven-deity arose, Dyaus, the god of the bright canopy of the day, before whose worship that of Varuna died away. We have the same god in the Zeus of the Greeks, the conqueror of his predecessor, Uranos. He again appears in the Teutonic Tiù, the god of light. The Odin of the Scandinavians, with the sun for his single eye, seems to be another heaven-deity. Again we have the heaven-god in his paternal aspect as the Dyaus-pitar of the Hindus, the Zeus Pater of the Greeks, the Jupiter of the Romans, — the kindly and beneficent progenitor of gods and men, the supreme parental deity of all that has life.

With the Hindus the sun was symbolized by a later deity, the golden-haired Indra, the god of light, whose arrows were each hundred-pointed and thousand-feathered. With the lightning for his beard, and brandishing a golden whip, he drove his flaming chariot across the heavens. The rains and the harvest were his gifts to mankind, while the demons which threatened the human race found in him a terrible foe. In Balder the Beautiful, the lord of light of the Teutons, we discover the Sun-god again, dying yearly at the winter solstice by the hand of the blind god Hödr, the demon of darkness, and rising again in his beauty as the shining summer returns.

But we cannot here attempt to name the interminable list of deities of the later Aryan worship, many of them, particularly in Greek mythology, borrowed from neighboring nations, and fitted, often very awkwardly, into the Olympian court of the Hellenic gods. It will suffice to say that this ancient system of worship is preserved to us in its most archaic integrity in the Vedas, — the work which holds the oldest recorded thoughts of man on natural phenomena. In it we have the deific host as the Devas, the shining ones; the dawn as Ushas, the bright, loving, gentle, white, and beautiful; the deities all simple in their attributes, and without the wide garment of myth that afterward enfolded them, — plainly the elements half transformed into the immortals.[1] We find ourselves here

[1] A striking instance exists in the story of Agni, the Fire-god of the Hindus. The Vedas tell us that two sticks were the parents of this deity, who was no sooner born than he turned upon and devoured them. Here is the original method of obtaining fire by the friction of two sticks transparently displayed. Yet Agni soon became one of the mightiest of the gods. He grew rapidly from his humble origin, flaming upward. as it were, from earth to heaven.

but a step beyond the archaic Aryan stage, in which these deities were yet clearly the powers of earth, air, and sky, and in which each was, for the time, the supreme being to his worshipper. Their deities had not yet been specialized as we find them later among the Greeks.

As the branches of the Aryan race left their primeval home and sought new lands of residence afar, certain highly interesting modifications came over their systems of worship, to which some attention is requisite. We do not refer to the expansion of their simple ideas of the deific attributes of natural phenomena into the splendid phantasmagoria of mythology, but to the characteristics of their religious organization. In this there was a marked difference between the eastern and the western Aryans. With the eastern branch the national or mythologic worship rose into supremacy, the priesthood became a powerful body, and the people fell under that dominion of priestcraft which has ever been such an opponent of human liberty. This was particularly the case with the Hindu tribes, over whom the priests gained an extraordinary predominance, unequalled in the history of any other people. The Hindu nation is one without great kings or great heroes. Its only great men are the lawgivers, the founders of systems, the priests of the race. When the tribes first marched to victory over the aborigines of India it was with the priests at their head. The Vedas are the record of the stirring hymns of praise or invocation with which these priestly warriors led their soul-stirred hosts. And when the Hindus sank to rest upon their conquered territory it was under the dominion of the priests. No great warrior led them to new victories, no powerful kingdom-maker welded the scattered bands into a nation, no earnest thinker wrote the history of the people. It was the history of the gods, not that of man, with which their thinkers were concerned; and we have grand systems of religious philosophy instead of a record of the mighty doings of man. The story of Hindu civilization is a phenomenon without parallel upon the earth.

The story of the Persians begins under conditions strikingly similar to that of the Hindus. Here, too, we behold a people marching to conquest with a priestly leader at their head. The great figure of Zoroaster dwarfs all the heroes of the sword. And their antique literature is religion, not history. It yields us only the outlines of that Zoroastrian system of faith and philosophy which was gradually filled up by priestly successors. But the location of the Persians forced them into a very different channel of history from that pursued by the Hindus. Instead of the hot, moist, enervating lowlands of the Indus and the Ganges, so favorable to the growth of superstitious belief in the divine power of the elements, they inhabited the bleak and inspiriting highlands of Iran. And the trumpet-blast of war rang everywhere around them, forcing them into battle for self-defence, and finally rousing them to victorious aggression. Great warriors and kings arose. The history of man began, and that of the gods ceased to be written. Yet to the late days of the empire the priesthood continued a powerful body, and, in alliance with the Throne, aided strongly in the subjection of the people.

If now we examine the religious history of the western Aryans a different phenomenon appears. In none of the western branches did a powerful and controlling priesthood arise, with the possible exception of the Celtic, in which the shadowy group of the Druids stands out with a prominence not attained by the priesthood of the Teutons, Greeks, or Italians As for the early history of the Slavs, we are utterly in the dark; but there is no trace of a priestly establishment, and but faint indication of the existence of a mythology. In the religious, as in every other respect, the home-staying Slavs seem most fully to have preserved the antique Aryan system, their creed remaining that of worship of the ancestral gods of the house and the clan, while mythology with them failed to advance beyond its elementary stage.

With the Greeks a rich and varied mythology arose, and an active public worship of the gods of the whole people emerged. Yet it never attained dominance over the humbler house-worship. The priesthood always remained an obscure body, without power in Grecian history, or control over the Hellenic people. The prevailing rites were those of the clan, not those of the nation. The literature was largely devoted to the gods, but it was almost void of deific philosophy. It dealt with the elemental deities in a somewhat playful spirit, humanized instead of spiritualized them, and wrought the mythical stories of their lives into the neat embellishments of poetry, not into the ground-work of vast theological philosophies. The gods of mythology were brought down to earth, looked squarely in the face by thinking men, laughed at, and dismissed. The whole fabric of myth and fable fell prostrate in splendid disarray, its rich fragments only to be used thereafter as poetic simile and metaphor. The worship of the ancestral spirits alone survived, while the thinking men of Greece set themselves to work to devise a secular philosophy of the universe. And Greece moved with unyielding steadiness toward democracy, largely through the lack of a priestly control of the public mind which usurpers could seize and wield.

In Rome priestcraft stood at no higher level than in Greece. The Roman people were from the first deficient in imagination, and mythology there attained but a stunted growth. The house and clan worship, on the contrary, shows itself more prominently than in Greece. We find traces of it everywhere in Roman history, as when Coriolanus, deserting Rome, seats himself by the hearth of his Volscian foe, and claims the protection, not of the Latin Jupiter, but of the hearth-spirit of the household he has entered. Even when the literature of Greece invaded Rome, and was imitated with all the fervor of the Roman mind, its mythologic feature obtained no special prominence; while the gods of the Roman mythology always remained vague and unspecialized, and little developed from their antique Aryan form. Priestcraft, in consequence, never gained any footing of power in Rome. The system of public worship was, indeed, mainly reduced to a phase of Shamanism, augury and divination replacing the creation of great religious ideas, which elsewhere ruled the minds of men. Thus in the development of the Roman State, religion never enters as an important political element. We perceive only a steady struggle between the democracy and the aristocracy, fought with secular weapons alone, with the growing supremacy of the democracy; until the inordinately powerful element of the army overthrew the whole ancient fabric of the State, and replaced it with a military despotism.

Teutonic history, so far as we are acquainted with it, tells the same story. There was plenty of imaginative fervor, and mythology gained very considerable development; yet but faint traces of a priesthood have survived. Possibly the worship of the household and the clan dwarfed that of the elemental deities. When the Teutons march to victory it is not with a priest at their head, nor even by the side of their military chief. No such figure makes its appearance, and the only Teutonic hero is the wielder of the sword. It was doubtless principally due to this reason that Christianity made such rapid progress with the Teutonic tribes. There was no one with a strong interest in preserving the mythologic faith, no one to control the tribes in matters of belief, no earnest clinging to the deities of mythology. The tribemen vaguely dreaded the vast gods of the elements, but their main worship was paid to the deities of the household, on whom alone their affections were centred. This private worship was too deeply ingrained to be eradicated except by slow degrees; but the weakly held mythologic faith was suffered to be replaced by the Christian creed with an ease that would appear frivolous did it not prove how shallow an impression mythology had made upon the Teutonic mind.

If we examine the early legend and fable of the several Aryan branches, an interesting illustration of their difference in religious condition appears. The ancient Hindu tradition has nothing to do with man. Only the gods appear in it, and its supernaturalism is wildly extravagant in character. Man is a creature not worthy to be named in a universe which contains the gods. Ancient Greek tradition tells a widely different story. In this, man is the central figure. The gods are present, it is true, and there is no lack of supernaturalism; but heroic man is their equal rather than their slave. He is displayed in steady struggle against the terrible powers of Nature, and in combat even with the Olympian deities. He is usually overcome and punished, yet he always retains something of the heroic; and the most striking figure in Greek mythology is that of Prometheus, the defender of man against the gods, terribly punished, yet eternally unsubmissive, and hurling threats from his rock of torture against Zeus, his deific foe. Nor are the gods always the victors. In the pages of Homer we find heroes daring to wound the gods, and escaping punishment for the impious deed.

If now we come to the antique legend of Rome it is to find the gods utterly forgotten, and man alone the subject of thought. It is admitted that the so-called history of ancient Rome is a tissue of fable; yet it long held its own as history from the fact that it dealt solely with human deeds. It is almost devoid of the supernatural. The gods hardly enter as agents. The old Roman saw only his hearth-spirits, or but vaguely beheld the elemental deities of ancient Arya. His imagination dealt solely with man and his deeds, in a series of stories that are sober history as compared with the exploits of the Greek heroes, and that breathe the most rigid spirit of the practical, as compared with the exuberantly fanciful Hindu conceptions.

This lack of a powerful priestly organization in the history of the western Aryans is without a counterpart in the civilized nations of the earth, with the one exception of China. That it has had much to do with the strong tendency to democracy in these nations, as compared with the tendency to aristocratic government elsewhere, can scarcely be questioned when we remember how powerful a controlling agent is religion upon the mind of man, and how vigorous is the grasp of the ruler who can seize at once the spiritual and the temporal reins of dominion.

The facts here given of the slight hold upon the western Aryans of their system of national religion, and the lack of an organized and influential priesthood to develop the public worship and to create a strong sentiment in its favor, are of interest for a reason above briefly adverted to. No bulwark existed against the inflow of a foreign system of belief, and we cannot be surprised at the rapid progress of Christianity. Rome was a fallow field to the seed of foreign religious thought. Its native faith was but feebly held, and we behold successively the Persian, Egyptian, and Christian creeds making their way into the Imperial City, with scarcely a word of protest or opposition, until the political danger from Christianity roused the dread of the Emperors and gave rise to spasmodic persecutions. Not a word of appeal for the old gods comes from the priests of Rome.

In Greece something similar appears. The systems of the philosophers there replaced the figments of mythology, and the opposition to this philosophy came from the conservative class of the people rather than from the priests. The after opposition to Christianity came from the adherents of the philosophers, with their proud admiration of the greatness of Greek thought. Mythology in Greece was dead before Christianity arose. Among the Teutonic clans the opposition to Christianity was nothing stronger than a vague distrust of strange gods. The voice of a chief in favor of the new faith carried with it his whole body of followers, who threw off their mythologic belief as easily as they might have discarded an ill-fitting cloak. No priest raised his voice in favor of the old gods. The hearth-spirits were as yet left to the people, and these were the only deities which had a hold upon their hearts. This phenomenon is singularly contrasted to the persistence with which the same tribes afterward clung to the slightest shades of sectarian Christianity. Instead of being without a priesthood, they had now come under the control of the most completely organized priesthood in human history.

VII.

THE COURSE OF POLITICAL DEVELOPMENT.

THE political organization of the ancient Aryans is one of the most interesting features in the whole history of human institutions. It has had an extraordinary influence upon the development of modern civilization, its basic conditions having maintained themselves with a remarkable persistence through long eras of tyranny and oppression. Finally, in the government of the United States we have what is in many respects a survival of the government of ancient Arya, so far as the simple conditions of the antique tribe can be brought into analogy with the complexity of relations in the modern nation. For in the Republic of the United States we possess a system of local self-government ranging upward through the family, the township or ward, the city or county, and the State, to the nation, with its general supervisory power over all below it. This is a close counterpart of the family, the village, clan, or gens, the tribe, and the confederacy of the ancient Aryans, each with its self-government in all that immediately concerned itself. It is the system of non-centralization, as opposed to the centralization which forms the basic feature of despotic government. In religion the same phenomenon appears. There was no State religion in ancient Arya, and there is none in modern America. The religion of the household or of the clan ruled in the one, as that of the person or of the sect does in the other. In despotic government, on the contrary, a centralized or State religion is an essential feature, and few tyrannies have been established without its aid.

The development of human institutions has been very little considered from this point of view; and before examining the Aryan system particularly, a brief comparison of this with the other systems of civilized mankind is of importance. Such a comparison will reveal features in the Aryan organization differing from those of any other family of mankind, and show clearly that ancient Arya was the true cradle of human liberty. Yet it will show at the same time that Arya was by no means the cradle of human civilization. Despite the very evident intellectual superiority of the Aryan race, its institutions acted as a strong preventive to political progress; and but for the activity of external agencies, and of influences at variance with its democratic organization, the Aryan peoples of to-day might be in the same state of stagnation that we find in the village communities of Russia and India.

In reviewing the early organization of human society, wherever advanced beyond the savage state, a remarkable uniformity makes itself apparent, indicating that the social and political conditions of mankind unfolded under the unconscious action of general laws, on the same principle that appears in the development of languages. Yet as human language, after pursuing the same course up to a certain level of unfoldment, diverged from this point into several different channels, so in the development of institutions a like phenomenon is manifest. Our purpose here is very briefly to glance at these lines of divergence.

The primal condition of man was undoubtedly a social

one. The lowest savages were combined in groups for various purposes. One of these was that aggregated for defence. A second was the family group, — probably definitely and firmly organized only at a late date. A third was the group for religious observance, — yet later in its concrete organization. Eventually these three groups appear to have become concentrated into one, that of the family. The family, with its secondary expansion into the community of kinsmen, became at once the social, the political, the religious, and the military group of mankind. Such is the condition of developing man everywhere that we can perceive him after he has advanced from the savage into the barbaric stage of culture. The family idea becomes the ruling principle in every interest of the tribe.

Early history, however, reveals to us two distinct stages in this unfoldment, — that of the patriarchal group, and that of the clan group; the latter an important step of advance beyond the former. The patriarchal system is that of Asia and northern Africa; the clan system that of Aryan Europe and North America. The desert was the native home of the patriarchal group. In the broad and barren steppes of northern Asia, and the great sandy plains of Arabia and northern Africa, the pastoral nomadic habit naturally persisted, agriculture in its faint first efforts remaining secondary to the interests of the wandering shepherd tribes. Communism reigned supreme. The flocks were the property of the tribe as a whole. Scarcely any individual property existed. The narrow confines of the tent, and the necessity of frequent movement, prevented the accumulation of any large amount of household treasures. Politically a like communism prevailed. There was no clear line of family demarcation. Each community was a group of kindred, and was under the leadership of the patriarchal representative of the remote ancestor of the tribe. But this leadership was by no means an absolute control. The separate families declared themselves sufficiently to form an assembly of freemen, not nearly so distinctly formulated as that of the Aryans, yet with a proud sense of personal independence, and a voice in the management of tribal concerns. The organization, however, was that of an army, with hereditary right in its leader, and subordination to his authority in all warlike affairs.

Religion was similarly communistic. We find no trace of any well-defined family worship, though there is evidence that a tribal ancestral worship prevailed. But combined with this was Shamanism, — a system of demon worship, in which incantation was the prevailing rite. Sorcery ruled as the main form of religion alike with the Mongolian tribes, the antique Semites, and the more barbarous tribes of North America. Very probably it had a strong footing also with the Aryans in their nomadic era, though it sunk into decadence at a later date. The only declared priesthood we can trace in this archaic stage of development is that of the Mongolian Shaman, the Babylonian sorcerer, and the American medicine-man or conjurer. Knavery undoubtedly had as much to do with their service as religion, and it must have been an easy task for the leader of the tribe to gain control of this venal priesthood, and thus add to the spiritual dignity which he possessed as the representative of the tribal ancestor. So far as we can trace, in every instance some degree of religious authority attached to his office.

All this may have nothing specially to do with the Aryans, but it is of importance from its decided contrast to the character of their organization and from the essential significance it bears in the history of human institutions. To the simplicity of the patriarchal system, indeed, we owe the original unfoldment of human civilization. But it was a civilization in what is known as the Asiatic form, — an unprogressive absolutism. Such is the condition which existed in the three non-Aryan civilizations of the old world, those of China, Egypt, and Babylonia. They were all patriarchal despotisms.

As already said, the nomadic tribe is a regularly organized army. It has its arms, and great ability in their use. It has its ready-formed regiments and divisions in the major and minor groups of the tribe. It has its clan-leaders, and its patriarchal tribal head, to whom all its members are willingly subordinate. And it is accustomed to swift and long marches, in which it takes with it all its property and food. No link of attachment binds it to a locality. Migrations are among the common duties of life. There is nothing to hinder invasion of a country at a moment's notice, settlement upon the land in case of victory, or swift retreat and disappearance in the desert in case of defeat.

The indications are strong that to this facility of warlike migration and this military type of political organization we owe the establishment of the early empires. China is most distinctively a patriarchal empire. Despite its long settlement, its developed agriculture, its abundant literature, its complex industrial and social conditions, it remains to-day politically a patriarchism, — the simplest and most archaic of all governmental systems. The emperor is the father of the empire. The long continuance of his absolutism arises from the fact that he stands at the head of the ancestral religious system of the nation. Ancestral worship has continued the ruling faith of China, and the emperor is the high-priest of this worship, — the hereditary representative of the primal ancestor of the people. He has inherited both temporal and spiritual power, and the bodies and souls of his subjects are alike bound captive. Like the house-father of old, the officiating priest of the house-worship and the family despot, the Chinese emperor is the only intermedium between his national family and the heavenly powers. He is answerable only to the gods for his deeds, and it is sacrilege to question his command. It is interesting also, in considering the character of Chinese civilization, to find that the ancient Shamanism still prevails. No developed elemental worship has been devised, all efforts to establish a philosophic faith have failed with the people at large, and the Taoism of to-day is undisguised sorcery. Yet it is probable that the Chinese empire arose ere the primitive ancestor-worship had been to any great extent superseded by the Mongolian Shamanism of to-day. In every feature of its organization, language, and belief, the archaic condition of mankind has persisted in China. This is largely due to the almost utter lack of imagination in its people; and the only civilized progress it displays is in devices for the practical needs of man, and in moral apothegms of the same tendency. The Chinese empire is the utmost unfoldment of the purely practical mentality of the Mongolian race.

In the early stages of the Egyptian monarchy we can somewhat vaguely perceive indications of a closely similar organization. The Pharaoh was the high-priest of his people, to whom he likewise bore a paternal relation.

There seems little reason to doubt that this empire was the outgrowth of a pastoral condition of society, that the emperor was the development of the original patriarch, and that his godlike dignity and absolute power arose from his being at the head of the ancestor-worship of the people, the hereditary representative of the primal ancestor. In early Egypt as in early China the absolutism of the emperor was not complete. There are indications of a tribal division of the people, and of the existence of a nobility with political powers. But patriarchism in its very nature tends to absolutism, and in both cases a complete subordination, alike of nobles and people, to the sacred father and emperor eventually succeeded. Religiously, however, Egypt developed far beyond China. Its people were of the highly imaginative Melanochroic race, and they devised a complex system of mythology, with a powerful priesthood, at whose head the emperor stood supreme. He was chief priest as well as sole ruler of the nation. As in China, he governed his people in body and soul.

Babylonia yields similar indications, though its organization is more obscure. Its earliest traceable religious system is a Shamanism, a highly developed sorcery. Upon this, however, arose a nature-worship, a somewhat complicated series of elemental gods. In regard to its governmental idea we are greatly in the dark. But its emergence in the heart of a pastoral region inhabited by patriarchal tribes, its absolutism, and the sacred or godlike character which plainly attaches to the later monarchs of Babylonia and Assyria, strongly indicate that it was a development of the patriarchal system.

It is singular and interesting to find that the archaic civilizations of mankind all apparently rose from the pastoral phase of society, — the simplest and most primitive method under which great bodies of men could be organized into national groups. Materially they all made great and highly important progress. Politically they remained almost stagnant. The simplicity of their system clung to them throughout, and absolutism continued a necessary phase of their national organization. The people submitted without a struggle, because their souls were bound in the same fetters that confined their bodies.

We may briefly advert to yet another national development of the pastoral tribes, from the interesting evidence to be gleaned from its literary remains and its present belief. The Hebrew people had distinctively a patriarchal organization, and their religious ideas present traces of ancestor-worship. Abraham was and is looked upon as the father of the race, its remote ancestor. It is not Abraham, however, but the god of Abraham, or rather a compound of this deity with the god of Moses, that is worshipped to-day by the Jews. The indication is strong that this special god of the Hebrew patriarch, the family god of Abraham, with whom he conversed and held personal relations, represented an ancestral divinity. The particular Jehovah of the Hebrews was the *Jahveh* of Moses, the family god of the Mosaic clan, as is clearly indicated in the Biblical narrative. He expanded with the growth of the Hebrew intellect into the supreme ruler of heaven and earth, yet to a very late day the Hebrews regarded him as the special deity of their race, their patriarchal divinity.

Coming now to the consideration of the American tribes, it is of high interest to perceive that they possessed the same type of family organization as that of Asia and

Europe, and that in this respect they were considerably advanced beyond the patriarchal system, and closely approached, though they did not quite reach, the clan type of the Aryans. Great differences in this respect, however, prevailed in different parts of America, some tribes being much more advanced than others. The barbarian tribes of North America, usually classed as in the savage hunting stage, yet really to a considerable extent settled and agricultural in condition, were organized on a definite clan-system, — a compound of kindred families like that of the Aryan village. This Indian organization, while closely resembling, differed in some important respects from the Aryan system. It was, indeed, intermediate between the patriarchal and the clan system, and represented an interesting phase in the natural development of human institutions.

Communism prevailed to a greater extent than with the Aryans. Not only land communism, but household communism existed with many of the tribes, and the isolation of the household and the tyranny of the house-father, so marked in the Aryan organization, does not appear in the Indian. Among the Iroquois of the North several families inhabited the same dwelling, with little separation of household rights; and in the case of the Pueblo Indians of New Mexico, whole tribes, numbering several thousands of individuals, are still found dwelling in single great habitations. With these tribes there is no division of the landed property, and in this respect their organization is distinctly patriarchal.

With the Indians of the southern United States, however, the Creek confederacy and the neighboring tribes, whose habits were much more agricultural than in the case of the northern tribes, an interesting advance in social and industrial conditions is indicated, their organization very closely approaching that of the Aryan village. Here the households were separate; and while the soil was common property, each family cultivated a separate portion of it, and was sustained in its claim to the use and products of this family field. In one respect only did the industrial organization differ from that of the Aryans. Each family, while controlling the produce of its own field and its own labor, was obliged to place a defined portion of the product in a village storehouse, whose stores were laid up for the good of the whole community. Hunters were also obliged to place there a portion of their game. This provident institution, resembling that of whose existence in Egypt we have evidence in the scriptural story of Joseph, constituted a form of taxation for the public good, and seems to indicate an advance in political conditions beyond the Aryan community, in which no such custom existed. In reality, however, it signifies a lower stage of development. It was a remnant of the general communism of the patriarchal stage of association, and one which seems to have worked adversely to the interests of American liberty.

This industrial condition extended farther north than would be imagined from what is generally known of Indian history. Historians of Virginia and Maryland state that the Indians of those localities had the custom of dividing their lands into family lots, and possessed common storehouses, in which a portion of the food had to be placed, under control of the sachem, whose power was to some degree absolute.

This brings us to a consideration of the political organization of the Indian tribes. It must be borne in mind, however, that in the Indian, as in the Aryan community, there was no such definite organization as is produced by a body of written laws. Custom was the only law of these communities, and there was doubtless considerable variation between different tribes. Yet the general principle of organization was everywhere the same. The system was an elastic one, which might stretch considerably, but could not easily break.

One marked feature of the Indian organization was the existence of two sets of officers, with definitely separated functions. These were the sachems and the chiefs, — the former distinctively peace-officers, the latter the leaders in war. These officers were elected; and in the elections it is of interest to find that the women of the clan had a vote as well as the men. Woman-suffrage is apparently a very old institution on American soil. The principle of choice of these two sets of officers, however, was very different. The war-chiefs were elected for personal valor, and there might be several of them in the clan. The sachemship alone was a hereditary office, and needed to be permanently filled; the new incumbent being usually, though not necessarily, chosen from the family of the deceased sachem, and perhaps vaguely representing the clan ancestor. The government of the clan was in the hands of all its adult members, male and female; while the tribe, made up of a number of clans, was governed by a council composed of the sachems and chiefs, and the confederacy, where such existed, by a council of the sachems of its constituent tribes.

No such definite arrangement existed in the Aryan clan. The principal chief there also probably had a hereditary claim to his office; but he was not distinctively a peace-officer, like the sachem, but a leader in war, and the council of freemen formed the executive body in matters of peace. His power was not distinctly marked off from that of chiefs chosen for personal valor or warlike ability only, and in time the distinction may have become wholly lost; the ancestral claim of the chief, which was never very strong, vanishing completely.

The Indian organization indicates an intermediate condition between the patriarchal and the Aryan village community. In the sachem we have the patriarch, shorn of some of his powers, yet not reduced to the mere war-leader of the Aryan clan. One important remnant of his old power existed in his control of the public storehouse. As the latter appears to represent a partial survival of the original general communism of the patriarchal tribe, so the control of it by the sachem represents the original control by the patriarch of all the wealth of the tribe. In neither case was this an ownership; it was simply a control for the good of the community. The mico — or sachem — of the Creek communities had no claim to the treasures in the storehouse, but had complete control over them. These had assumed the shape of a general taxation for the public good, and he was the general executive officer of the community, with a considerable degree of arbitrary power in his administration. His government, however, was controlled by the village council, which met to discuss every question of equity and to try every case of crime.

There was one further feature of interest in the Indian organization to which we must now advert, — that of their religious conceptions. Among the savage tribes of the North, Shamanism appears to have been the prevalent faith, and sorcery the prevalent practice. The medicine-man was the religious dignitary, his influence over the tribe being that of fear rather than of awe and spiritual dignity. The worship of ancestors is not indicated, while no elevated religious conceptions are displayed. A vague polytheism seems to have existed, with belief in a "Great Spirit" and a series of lesser gods; yet this was undefined, and nothing that can be called a mythology had arisen.

Among the southern tribes, however, a very different state of religious belief prevailed. They possessed a mythological religious faith, with the sun for supreme deity, while their worship was conducted with all the ostentation of temples, high-priest, and a considerable priestly establishment. The democratic religious system of the Aryans did not exist among them. Their religion was aristocratic in tendency, had a vigorous influence over the minds of the people, and afforded a ready instrument for their subjection. While, indeed, there was a high-priest, the mico was the real head of the religious hierarchy, and added to his temporal influence the power arising from spiritual dignity. The patriarchal position of spiritual head of the tribe adhered to him, though the ancestral worship, to which he may have owed his original religious authority, had vanished.

The final outcome of this condition of affairs appears in a tribe to the west of the Creeks, the Natchez. The government of this tribe was an absolute tyranny, the power of the ruler being based on his religious dignity. He had become "The Sun," a god on earth, and the people were slaves to his will. There was an intermediate class of nobles, — perhaps the remnant of the former council; but "The Sun," the earthly representative of the supreme deity, was absolute over the entire community. The organization of this tribe presented some other interesting features, which we have not space to describe, but which were in conformity with the principles above indicated. It constituted a patriarchal despotism in close conformity with those of Asia.[1]

As to the origin of this peculiar state of government and religion among the southern Indians, so different in some respects from those of the wild tribes of the North, we have much warrant to consider it a survival of the organization of that vanished race known as the "Mound-Builders," which at one time occupied the whole valley of the Mississippi and its tributaries, but which seems to have been dispossessed by the bordering savage tribes, partly annihilated, and perhaps partly crowded back into the southern range of States, where it left its descendants in the Natchez, the Creeks, and others of the southern tribes.

A brief glance at the Indian civilizations of Mexico and Peru will lead us to conclusions like those above reached. In Mexico absolutism was not fully declared. The Montezuma, the spiritual and temporal superior, was controlled by a council, — the survival of the old tribal assembly. Yet he was rapidly advancing toward complete absolutism at the period of the Spanish invasion. The storehouse of the northern tribes was here represented by an extended system of taxation in kind, over which he had full control, while his position as supreme pontiff gave him an influence

[1] For fuller information concerning these interesting institutions of the American Indians, the reader may be referred to Jones's "Antiquities of the Southern Indians," in which the organization of the Creeks and Natchez is fully described, and Morgan's "Ancient Society," which gives valuable information in regard to the Iroquois confederacy and the general governmental relations of the Indian tribes.

which threatened to overthrow the feudal power of the nobility.

In Peru existed an absolutism as entire as that we have seen among the Natchez. The Inca was autocratic both in religion and in government. He was the descendant of the gods and a god himself, whose mandate none dared question. A nobility existed, but it was a nobility without authority, except such as emanated from the Inca. The land and all its products were at his command. Village establishments existed, with division of family lots; but a large section of the land belonged to the Inca and the church, and was worked by the people for their benefit. The product of the royal and Church lands was stored in great magazines, the direct counterpart of the storehouse of the North, since their contents were held for the good of the whole community, though subject to the Inca's absolute control. It was unquestionably the spiritual dignity of the emperor, in all the civilizations named, that caused the entire submission of the people to his will, and that subordinated the nobility as fully in the peaceful empire of China as in the warlike empire of Peru. It is surprising to find so close a conformity existing in the principles of Indian organization throughout the wide range of North and South America. Nothing could show more clearly the supreme influence of natural law over the development of human institutions.

Yet there was another agency necessary to the production of the final effect, of the utmost importance in this connection, — that of war. Much as human hostility and bloodshed may be deprecated, the fact is unquestionable that to it we owe all accelerated steps of human development. Even in this advanced age, war was necessary for the rapid annihilation of slavery in America, and has yielded within a few years a degree of political and industrial progress which otherwise might have taken centuries. In savage and barbarian communities it is the all-essential element of progress. The conservative clinging to old conditions and institutions, which is yet vigorous in modern nations, was a hundredfold more so in the early stages of human progress, and war was the only agent sufficiently radical and energetic to overthrow old ideas and customs, and reorganize society on a new basis.

We can here but briefly glance at its general effects. One of the first and most important of these is to increase the authority of a successful chief and to bring new tribes under his control, either as allies or as conquered subjects. The equality of the freemen of antique communities was rudely broken into in states of war. The patriarchal tribe at once became an army, and was subjected to army discipline, which included autocratic power in its chief. On regaining a state of peace this absolutism of the chief over his followers did not entirely vanish, while it remained strong over the conquered tribes. The general effects of war at that stage of human culture were the following: The principle of human equality was dissipated, and society divided into classes, composed of the principal chief, or king; the secondary chiefs, or nobles; the freemen, of the conquering tribes; and the subjects, or slaves, of the conquered tribes. Some such division seems to have been an inevitable consequence of continued war, and appears as well in the development of Aryan as of patriarchal institutions; and in every instance some condition approximating to that of feudalism seems to have emerged. It existed in Mexico at the era of the Spanish conquest. It had very probably existed in Peru at an earlier period. Indications of its existence in Egypt and China appear. And in the empire of Japan it continued in existence until very recently. But in every instance it has disappeared under the growing power of the king. In Egypt and China we perceive the monarch of a province gradually extending his authority over the whole country by successful war. A similar phenomenon appears in Mexico and Peru. In every such case the chiefs of the conquered tribes became the nobles of the new empire, with some remnant of authority. But in all the cases mentioned, the power of the nobles gradually vanished, and that of the monarch became absolute.

This phenomenon was undoubtedly due to the religious position of the monarch of these patriarchal empires. Where the body would have vigorously resisted, the soul sank in powerless slavery. In every one of the four empires named, the emperor was supreme pontiff, the head of the religious establishment, the son and representative of the gods, and the connecting link between earth and heaven. It was the recognition by the people of this spiritual dignity in the emperor, their superstitious awe, and the moral support which they gave him in his encroachments upon their liberties, that rendered the resistance of the nobility unavailing. Step by step they sank until they became ciphers in the state, with nothing but a title to distinguish them from the people. This is the condition which exists to-day in China, where the nobility and the people stand on an equal footing in respect to the authority of the emperor.

A highly interesting recent case in point is that of Japan. Our early historical knowledge of that empire reveals a strong feudal nobility, with a spiritual emperor of reduced authority. A powerful chief, the Tycoon, or Shogun, through the influence of his position as head of the army, succeeded in robbing the Mikado of nearly all his temporal authority, and taking the reins of power into his own hands, leaving to the titular emperor little more than his title. But the people remained spiritual subjects of the Mikado, their souls in submission to him, while only their bodies were governed by the Tycoon. This powerful basal support has enabled the spiritual emperor, during the disturbances caused by the forced opening of Japan to foreign intercourse, to overthrow his rival, bring to an end the feudal institution, and make himself unquestioned autocrat of Japan. After a long interregnum patriarchism has there reached its inevitable result, — that of the spiritual and temporal absolutism of the emperor. The patriarchal empire, while naturally the simplest in organization and the easiest established, was one that tended inevitably to autocracy and subjection. For the establishment of liberty in civilization the growth of a widely different system was necessary. And this we find in the Aryan organization.

It is of high interest to perceive the great degree of conformity that existed in the unconscious development of human institutions. Patriarchism seems to have always evolved as the first stage beyond savagery. We find it widely disseminated in Asia and northeastern Africa, with its final culmination in despotic governments. Throughout America society, under the influence of agricultural industries, had advanced a stage beyond patriarchism. Yet the civilizations there arising tended inevitably toward absolutism. For the establishment of democratic institutions a further step of advance in barbarian organization was necessary; this step forward we have next to consider.

The description above given of the political characteristics of the other barbarian and civilizing tribes of mankind is of importance from their marked contrast to the Aryan condition, and as indicating the special features to which we owe the Aryan type of civilization. This type, we may say here, was overturned in two of the Aryan empires,—the Persian and the Macedonian, — which deliberately adopted the Oriental system, and maintained it by the power of the sword and by the fact that their subjects were largely Semitic and long accustomed to despotic rule. It was partly overturned in the Roman empire, as a result of continual war and the subjection of the State to the army and its chief, though the senate of Rome kept intact the principle of the Aryan assembly to the last, and the emperors never succeeded in their efforts to attain spiritual authority and to command the worship of their people. In no other Aryan nation has the effort to kill out the spirit of ancient Arya attained any marked success. Democracy and decentralization have unyieldingly opposed the efforts of aristocracy and centralization.

It is singular within what definite limits human progress has been confined. In every case of development beyond the savage state we find the family organization gradually unfolding into patriarchism. In two families of mankind, the Asiatic Mongolian and the Semitic, progress stopped at this point, in conformity with the pastoral character of their industries, and patriarchal civilizations arose, their early development being due to the simplicity of their system, and the ease and completeness with which it permitted the control, movement, and subordination of large bodies of men. In two other families, the American and the Aryan, development proceeded further as a result of the change from the nomadic pastoral to the agricultural condition, and produced the clan or village system; and it is remarkable, considering the impossibility of intercourse between these two races, how closely their organizations resembled each other. In both we find the village system, the democratic assembly and election of officers, the combination of families into clans, of clans into tribes, of tribes into confederacies. In both, the organization of the people was personal, not territorial. In both, communism in landed property prevailed. In both, patriarchism existed to the extent that a certain family in each clan was considered of purest descent, and usually furnished the clan rulers. Yet, as we have shown, the American system retained the principle of communism in a much greater degree than the Aryan, and this communism extended to religion. The democratic system of Aryan worship had not appeared, the sachem was at the head of the spiritual establishment of the more civilized tribes, and he became the representative of the Sun, as the Egyptian Pharaoh did of Osiris, and the Chinese emperor of the vaguely defined heaven deity, while absolutism appeared as a direct consequence of this spiritual autocracy.

The distinctiveness of the Aryan organization lay in its complete development of the clan-system, its suppression of community in property beyond partial land-communism, and its almost complete suppression of religious communism. In ancient Arya each house was a temple, each hearthstone an altar, each house-father a priest, each family a congregation, with its private deity and its private ritual of worship. Some minor degree of communism existed in the general ancestor-worship of the clan and in the less influential worship of the elemental deities; but the hearth-spirit seems to have been the favorite god of the Aryan, and a remarkable decentralization in religion prevailed. No people has ever existed more free in soul from the reins of spiritual authority. The Aryan house-father was a freeman before the court of Heaven, as he was in the assembly of his tribe. It was impossible for any ruler to hold him fettered body and soul like the subject of an Oriental monarchy. Mentally he was in eternal rebellion against tyranny. And it is to this that we owe the political liberty of modern Europe and America. Yet the decentralized and democratic organization of the Aryans was strongly opposed to that concrete and definite association in large, settled masses which seems everywhere to have been a necessary preliminary to civilization. A considerable degree of political consolidation has everywhere preceded material progress, and to this the Aryan spirit was vigorously opposed. It is one of our purposes in this inquiry to trace how this opposition was overcome, and how the village community developed into the State.

We have already in previous sections described to some extent the Aryan tribal organization, — the political system which prevailed in ancient Arya, and of which indications appear in the early history of all the branches of the race. It is a problem of interest to trace the evolution of the family into the clan, of patriarchism into democracy. In the largely patriarchal Highland tribes of Scotland there existed minor groups of fifty or sixty clansmen, with a particular chief, to whom their first duty was due. This is analogous to the Slavonic house community, whose members range from ten to sixty in number. When grown too large, a swarming to found new families takes place. But this in itself does not break up the close patriarchal family relation. Two further steps are necessary to clanship, — the apportionment of a separate lot of land to each new family, and the development of a system of home worship.

This is what occurred in the Aryan clans, each of which was formed of a group of several families descended from a common ancestor and with a separate organization of its own. It was ruled by an assembly of the house-fathers; though this mode of government was gradually subordinated to that of the chief, elected by the assembly, but usually from a privileged family. It had its system of clan-worship, its common burial-place, and its common landed property. There was no occasion for any householder to make a will. The property-rights of a deceased member descended to his fellow-clansmen. No definite legislation existed. The clan was governed by a series of ancient customs, the growth of centuries of usage. The assembly was an executive, not a legislative body, though it seems to have legislated sufficiently to meet business exigencies not previously provided for. To these clan conditions must be added another of considerable importance, — that of the duty of common defence, common revenge, and common responsibility. Each clansman was bound to defend his fellows, to exact retribution, in money or blood, for injury to a fellow, and was himself responsible for any criminal act committed by a member of his clan. The whole clan of a murderer was held accountable for the murder, and blood-revenge might be taken upon any member of the offending clan. No true sense of individuality existed. Each clan was an individual, and the whole clan, or any part of it, was responsible for the acts of any of its members. On the other hand, damages awarded to any person for injury received, belonged not to him, but to his clan. It was the duty of each clan to restrain its members from crime, and this duty was accentuated by a general responsibility.

Though we cannot look into ancient Arya itself, we can perceive these conditions as they left their mark on subsequent Aryan law. In old Anglo-Saxon law, for instance, the duty of each clan to act as a police upon its members, its money responsibility for any crime committed by a member, and its equal share in damages awarded to a member, are clearly shown. But the traces of this custom have descended still lower, and may be found rather widely spread to-day in the system of the vendetta or blood-revenge, which exists among all half-civilized Aryan peoples. We know to what an extent it formerly prevailed in Corsica, from which point it still extends as far east as Afghanistan. In this custom it is the duty of every member of a family, one of whose near kindred has been murdered, to exact blood-revenge from any member of the murderer's family. The Southern United States were the seat of a well-developed vendetta system of this character in the *ante-bellum* days, and cases yet occasionally crop out to show that the spirit of antique Aryanism is yet alive in the benighted regions of this country.

As for the tribal combination of the Aryan clans, it is doubtful if it existed as a permanent group in ancient Arya; and the confederacy of tribes arose only under the influence of migration and warfare. It appeared among the Teutonic people only after they were forced into strong combinations by long conflict with Rome It may be further said of the clan-organization that it was vigorously maintained. None could leave it without permission from the council, and no new member could be admitted without a ceremony of initiation. The clan-council seems in some cases, or among certain tribes, to have been limited in number. Evidences exist of an ancient council of five in Greece, Rome, and Ireland. This limitation does not appear elsewhere. It should also be said that, in addition to the agriculturists, the clan contained hereditary artisans. Commercial pursuits, however, such as the business of the grain-dealer, do not seem to have been hereditary.

From what has been said, it will appear evident that the antique clan-organization was one of very great simplicity. There was nothing that could be called criminal law, though there were many rules of business procedure. There was no legislator and no executive. Each clan took on itself the duty of punishing crime against itself. It was not the duty either of chief or council to see that justice was done between persons. The council mainly concerned itself with the care of the common property and with the good of the clan as a whole. The chief was personally active only as a war-leader. He had no special duty or authority in peace. Of courts, laws against crime, or officers of justice, we have no indications. The family was under the autocratic control of the house-father. Revenge for wrong was the duty of the kindred of the injured person, who might exact damages in property or in kind. Injury from outside the clan it was the duty of every clansman to avenge.

The military system was as simple as the civil. The clan was the basal unit of the army, and marched to war under its chosen chief. A group of such clans, under a tribal chief, formed an army. Every freeman was a soldier. The military system existed ready formed in the civil. This is clearly indicated in the Celtic and the Teutonic warlike organizations; and an interesting evidence of the existence of a similar system in Greece is given in the Iliad, in which old Nestor tells Agamemnon to muster his men by *phyla*[1] and by *phratra*,[1] so that each clansman might support his fellows in the ranks. Of the early Roman system we are in ignorance.

Yet another survival of the ancient clan-system may be spoken of here, — that of the co-operative guild, or trade, which existed in Greece and Rome, in old Ireland, and was largely developed in Middle-Age Europe. A similar system exists in Russia to-day, where its development from the village community organization is very evident. In addition to the communistic guilds of workmen in the cities, many villages are arranged on the principle of communistic artisanship. We are told that there are Russian villages where only boots are made, others whose inhabitants are all smiths, and some, indeed, which contain only communistic beggars.

This review of the system of clanship as a political condition may be followed by a consideration of the later stages of growth in Aryan institutions. The clan-system in its purity was adapted only to a barbaric stage of society. Further development could take place only through the entrance of new elements into the situation. It may be said here, however, that in Attic Greece a vigorous republic was established, that differed in organization from the ancient tribal system in only one essential particular, — that of the replacement of family by territorial relations;

[1] Sub-divisions of the tribe.

and that the great republic of the United States is but an expansion of this idea. Communism has died out, the council is composed of elected representatives instead of the whole body of freemen, and men are grouped in territorial divisions instead of kindred groups; but with these exceptions the political system of the United States constitutes a direct development of the method of organization of our remotely prehistoric ancestors.

The clan-element which gave rise to the historic development of Aryan institutions was that of chieftainship. It was an element of individualism placed side by side with that of communism. It was an inevitable outcome of the situation, and one destined, with the aid of warlike aggression, to carry the Aryans far forward on the road of progress. To its evolution our attention must now be turned. In process of time the idea of kinship became more and more of a fiction in the Aryan clan. The family had its dependents, and in the warlike period its slaves and freedmen. The clan in like manner had its dependents, who after three generations of service acquired a hereditary right in the soil. The increase of this alien element exerted a very important influence upon the history of Greece and Rome, as we shall see further on. It will suffice here to say that the wealth and superior position of the chief enabled him to surround himself with a larger body of dependents than was possible to ordinary freemen. His estate was apparently an independent household, organized on the old patriarchal system, with its own lands, its own cattle, and its own group of slaves and laborers. It was a house community on a large scale. This state of affairs, if not originated, was certainly enhanced by war.

Nor was it alone the hereditary and the elected chiefs who acquired this special importance. Any one with warlike reputation enough to attract followers could gather around him a body of retainers, mainly composed of warlike youths who were ripe for battle. And there was no hindrance whatever to such a person separating from the village and starting an independent establishment. Over such retainers the chief acquired an authority like that of the house-father over the family. He was their absolute lord, to the power of life and death. They could leave his service if they wished, but were the subjects of his will while they remained. The tie of connection was a tie of honor, and its strength may be seen in the ardent devotion of the Teutonic and Celtic clansmen to the cause of their chief.

The incessant wars that prevailed during the period of migration added greatly to the power and influence of the chiefs. To those with hereditary title to their chieftainship were added those elected for their valor, and perhaps those who gained influence through their wealth and personal powers of attraction. Through the above-named influences the community gradually became divided into the three classes of nobles, freemen, and slaves. Not that the nobles had any political authority over the freemen, or could set aside the voice of the assembly; their dignity was solely personal. Yet war and conquest had their inevitable effect in adding to the inequality in wealth and power. The chief naturally seized the lion's share of the spoil, and used it to increase the number of his followers. And subject-villages became subordinate to him personally rather than to the clan. Over these he gained some degree of political authority and rights of taxation.

Step by step the ancient system became subverted, and a new system of individual authority established, as war gave the warrior precedence over the citizen. Indications of this growth of aristocracy can be seen in every branch of the Aryan race, from the Rajput nobility of India, to the chiefs of Greece, Rome, and Germany, and the so-called kings of Ireland.

Maine says of the Irish chiefs that though they formed to some extent a class apart, they stood in closer relation to the septs they presided over than to one another. There is some reason to believe that the tribal chief had gained a portion of the authority of the Druids, and acted as priest and judge as well as war-chief. The popular assembly, so powerful in Greece and Rome, had lost all judicial authority over the Irish Celts. Property was rapidly losing its communistic character. The chief claimed ownership of large individual tracts, as well as certain rights in the communal lands; villagers claimed to own the communal lots they had long cultivated; and a system of petty usurpation had set in, apparent to a greater or less degree in all Aryan regions, that threatened in time to completely overturn the old system of land-holding. To it, aided greatly by war and the seizure of large conquered estates, we owe the establishment of feudalism, — the natural outcome of Aryan communism and chieftainship.

The political development of Greece and Rome is of interest in this connection, as indicating one of the two natural methods of unfoldment of the Aryan system. It is the development due to the influences of city life as contrasted with that arising from the agricultural condition. Its purest display is that seen in Attica. Here we have to do with a sea-going commercial people, industrial in habit, except to the extent that necessity drove them to war. Into the active city that naturally arose under these conditions, aliens crowded from all sides. Yet the early form of government was strictly an organization of gentes, or clans, the old Aryan personal system which had held its own in the formation of the civic government. To the new conditions it quickly proved inadequate. The great influx of strangers, members of no gens, and jealously excluded from gentile privileges, in time brought the government into the hands of a few ancient families, who conducted it on the old clan-system, except to the extent that the chiefs of the gentes acquired political authority and replaced the ancient democratic by an autocratic rule. The growth of chieftainship can be clearly seen in the story of the Iliad, it being highly probable that the "kings" of old Greece had but the standing of tribal chiefs, with an authority augmented by the warlike subjection of neighboring clans and the adherence of alien dependents, while the voice of the assembly had become a mere agreement in the proposals of the chief.

Undoubtedly there was a strong pressure from the alien population of the city of Athens to gain a share of political rights, and as strong a determination of the gentes to hold the reins of power. It became more and more evident, as the difficulty grew more urgent, that some reform must be adopted, and several measures were proposed by influential chiefs or lawgivers. The first of this is a traditional one, ascribed to Theseus. He sought to consolidate the tribes into a nation, with one instead of many councils. He also attempted to divide the people into the three classes of nobles, husbandmen, and artisans.

This legendary division was found in existence in Attica in the seventh century B. C. But the gentile system of organization was in full vogue at that period. At a later date we find the people gradually overthrowing the usurped authority of their chiefs. The *basileus*, or king, lost his weak priestly authority, and was thenceforth called *archon*, or civil ruler. Later again this hereditary life-office was made elective, and limited to ten years. Finally it was made annual, and divided among nine archons. Thus the partly overthrown authority of the popular assembly was gradually resumed, and the will of the people became the law in Attica.

The second definite effort at political reform was that of Solon, who divided the people into classes on the basis of property. This, however, did not do away with the division into gentes. The assembly under his laws gained increased, or at least better defined, rights, and became an elective, a legislative, and to some extent a governing body. But the bottom of the difficulty was not touched by these reforms, and could not be while the gentile families held all power. The final reform was that made by Cleisthenes (509 B. C.). He divided the people on a strictly territorial basis, without regard to their ties of kindred. Abolishing the four ancient Ionic tribes, he formed ten new tribes, which included all the freemen of Attica. The territory was divided into a hundred demes or townships, care being taken that the demes of each tribe should not be adjacent. It was a distinct effort thoroughly to break up the old clan-system. Each citizen was required to register and to enroll his property in his own deme, without regard to his ties of kindred. Each deme had rights of self-government in local matters, while controlled in national matters by the decision of the State government. Under this institution arose the primal republic, the measure and model of all subsequent republican governments. This reform was undoubtedly made in response to the demand and sustained by the power of the alien people of Attica, who must now have been sufficiently numerous to defy the gentes.

It is of interest to find that the government of Rome, without any knowledge of what was taking place in Athens, passed through essentially similar steps of development. In fact, the formation of territorial government in Rome is claimed to have preceded its establishment in Athens. It was a natural and inevitable line of civic growth. The same difficulty arose in Rome as in Athens. The inflow of aliens brought a strong pressure to bear on the system of gentes. The aliens demanded a share in the government, which was resisted by the clansmen. The earliest effort at reform is traditionally ascribed to Numa, who is said to have classified the people according to their trades and professions. This failed to produce any definite effect, and the Romans were still divided into the patricians, the old gentile clans, with full control of government; their clients, or dependents; and the plebs, or commons, the new class of aliens, without a voice in political concerns.

To overcome the discord that arose from this state of affairs Servius Tullius (576–533 B.C.) instituted a reform closely similar to that of Cleisthenes. He divided the territory of Rome into townships or parishes, and the people into territorial tribes, which crossed the lines of the gentes. Each citizen had to enroll himself and his property in the city ward or the external township in which he resided. This monarch is also credited with the establishment of a new popular assembly, which abrogated that of the gentes, and admitted each freeman to a voice in the government. Unfortunately, in addition to this wise arrangement he made a second division on a property basis, — establishing five classes according to the amount of their respective property. This mischief-making scheme separated the people at once into an aristocracy and a commonalty on the line of wealth, and gave the impulse to a struggle that continued for centuries. In Rome, as in Greece, we find the people gradually rising in power, and the government becoming a more and more declared democracy, though the struggle was here a very bitter and protracted one. It was finally brought to an end by the inordinate growth of the army and of the power of its leaders, by whom a vigorous despotism was established.

In Greece, however, the power of the people grew rapidly, all aristocratic authority quickly disappeared, and a disposition manifested itself to combine the several minor states into a confederacy, with a general democratic government. The antique Aryan system was here expanding, under the strict influence of natural law, into an ancient counterpart of the modern United States. Unfortunately for the liberties of mankind, it was overthrown by the sword of Rome ere it had grown into self-sustaining strength. During these many changes the ancient gentes continued to exist as separate religious organizations; but their antique political and communal constitution utterly vanished.

In the political development of the Teutonic tribes widely different conditions appeared. Their industries continued agricultural, and their unfoldment was more strictly in the line of the village system. Territorial government remained subordinate to personal government. The powerful invasions by which the empire of Rome was overthrown, and new states founded on its ruins, naturally gave immense power to the chiefs, which was increased by the incessant wars that succeeded and continued for centuries. The original independent establishment of the chief expanded into the feudal manor, and the chief into the feudal lord. His power was absolute. The house-father was reproduced in the lord of the manor. Below him were the descending grades of wife and children, dependents and slaves, as in the Aryan family. Around him were his retainers, bound by ties of mutual honor and subject to his will. His relation to them was that of military superior and of chosen companion in arms. As for the constitution of the feudal state, with its successive ranks, each lower one being held as military subordinate to the higher, but each, from the lowest noble to the king, being free from any obligations beyond that of military duty, and being absolute lord of his own territorial establishment and his retainers, we have in it a direct expansion of the original Aryan system, with marvellously little change in principle. The Aryan village and tribe, with the chieftain and his dependents and retainers, and his rights of suzerainty over conquered villages, formed the direct though simplified prototype of the feudal state, with its more complex system of obligations and wider extension of authority.

In considering the development of the Aryan village-system into the modern European state we find an interesting illustration of the persistent force of archaic ideas. Ancient Arya, as we have seen, contained, side by side, a double system of government. The village was essentially a democracy. But beside, and perhaps to some extent over it, was the patriarchal establishment of the chief. In the development of the feudal state both these conditions persisted, and the subsequent national history of Europe has been mainly a struggle between them for precedence. The patriarchal establishment of the chief, being the simpler and more centralized, and being one to which war added strength, rose first to power, and in some states developed into a degree of absolutism, though its lack of control of the religious establishment prevented it from becoming completely autocratic. But the democratic idea, though slower in its development, never died out, nor did the subjection of the people ever extend beyond their bodies to their minds and souls. The eventual supremacy of democracy was inevitable. In every era of peace it gained vigor, and to the extent that peace became the prevailing rule its demands grew more energetic and its victories more decided. At present it has risen into complete ascendency in America, while in Europe absolutism is shrinking before its force, and must inevitably everywhere give way to the "government of the people by the people."

With a rapid review of the political development of human civilization, this chapter may close. As we have seen, in two regions of the world patriarchism gained absolute supremacy, democracy failed to develop, and three states were formed on this simple system of paternal and spiritual absolutism, — Egypt, Babylonia, and China. One only of these has persisted unto to-day, — that of China; and in it not a vestige of a democratic idea has ever made its appearance. In America the growth of democratic institutions made greater progress, though in the two civilizations that arose, the spiritual authority of the emperor enabled him to completely overthrow them in the one case, and seriously threaten them in the other.

In ancient Arya the political development of barbarism went farther. Democracy gained a marked development both in political and spiritual affairs; the growth of a priestly autocracy was checked by the system of individual worship; and the patriarchal authority of the chief lost much of its force. The principle of election grew upon that of heredity. In the development of every Aryan civilization differing conditions operated, though it is remarkable what persistency the ancient ideas everywhere displayed. It is not necessary here to review all the Aryan states separately. In only two of them the ancient Aryan ideas developed with little external interference. One of these we have already considered, — that of Greece, in which the development proceeded under civic and commercial influences. The other is that of England, in which the Teutonic agricultural influences mainly prevailed.

Of all the European States, that of Saxon England was least disturbed in its development by external forces. The Norman invasion for a time gave supremacy to patriarchism; but this gradually yielded again to the steady persistence of the democratic idea. The Aryan popular assembly held its own as the English parliament, and has, step by step, taken control of the government, until, finally, it has left to kingcraft only its name and its palace. Fortunately for European liberty, the priestly establishment which eventually arose remained definitely separate from that of the kings, and usually hostile to it. The bodies of Europeans have been ruled by the Throne, but never their souls. Thus it was impossible that they could be reduced to the slavery of the Oriental system. Every effort of the kings to seize spiritual authority has failed, the spirit of democracy has steadily grown, and the promise is that ere many centuries not a trace of absolutism will be left on European soil.

Aryan political evolution has everywhere followed the same general direction; but its rapidity has been greatly affected by the conditions of society. Under the civic institutions of Greece and Rome, democracy, territorial division of the people, and private ownership of land early appeared; while with the agricultural but warlike Teutons and Celts progress in this direction has been much slower; and among the agricultural, but peaceful and sluggish, Hindus and Slavs, the ancient conditions still in great part prevail. Yet in every case the general course of evolution has been the same, and but one final outcome can be expected to appear, — that of complete democracy. In the patriarchal empires of Asia, on the contrary, political evolution followed an exactly opposite course, and long ago reached its inevitable ultimate in complete absolutism. Political progress in these empires has long since ceased, and can only be resumed under the influence of Aryan ideas and a reversal of the governmental principle which has so long held supreme control.

VIII.

THE DEVELOPMENT OF LANGUAGE.

LANGUAGE formed the clew through whose aid modern research traversed the Aryan labyrinth,—that mysterious time-veiled region in which so many wonders lay concealed. It cannot, indeed, be doubted that even without the aid of language this hidden problem of the past would have been in part solved. We have already shown that the Aryans have much in common besides their speech. Their industrial relations, their political systems, their religious organization, their mythologies, their family conditions, form so many separate guides leading to the discovery of that remarkable ancient community. Nor is this all. As we shall show farther on, the modern Aryans have still other links of affinity, less direct, it is true, than those so far traced, yet adding to the strength of the demonstration, and enabling us still better to comprehend the conditions of that ancient and re-discovered community.

Yet, with all this, the fact remains that language offered the simplest and safest path into the hidden region, and that by comparison of words we have found out much concerning the modes of life in old Arya that otherwise must have remained forever unknown. This being the case, it becomes a part of our task to consider the character of the method of speech which has proved of such remarkable utility in the recovery of a valuable chapter of ancient history. It is known to differ in important particulars from all other types of human language, not so much in its words, — for there many accidental coincidences with other languages exist, — but in its structure, in that basic organism of thought which is clothed upon with speech as with a garment. Yet in order properly to understand these structural characteristics, it will be necessary briefly to review the several types of speech in use by the higher ranks of mankind. A comparison of these types will reveal, as all philologists admit, that the Aryan is the most highly developed method of speech, and the most flexible and capable of all the instruments of thought yet devised by mankind. In this respect, as in all the others noted, the Aryan in its original organization was superior to the other human races.

The types of speech in use by the barbarian and civilized peoples and nations are divided by philologists into four general classes, — the Isolating, the Agglutinative, the Incorporating, and the Inflectional; the last being separated into two sub-classes, the Semitic and the Aryan, which properly should be considered as distinct classes. Of these methods the isolating is usually viewed as the least progressed beyond what must have been the original mode of speech. It is the one in use by the most persistent of human civilizations, — the Chinese. In the language of China we seem to hear the voice of archaic man still speaking to us down the long vista of time. It is primitive, as everything in China is primitive. Yet through the aid of a series of expedients it has been adapted to the needs of a people of active literary tendencies.

Philologists are generally satisfied that man first spoke in monosyllables, each of which conveyed some generalized information. The sentence had not yet been devised, nor even the phrase; and language consisted of isolated exclamations, or root-words, each of which told its own story, while no endeavor was made to analyze the information conveyed into its component elements.

Yet this idea directly affiliates the language of primitive man with that of the lower animals. For the lower animals possess a language of root-sounds, each of which yields a vague and generalized information, or is indicative of some emotion. Ordinarily this language consists of very few sounds, though in certain cases it is more extended, and is capable of conveying some diversity of information. This is particularly the case with some of the birds. And it is usually a language of vowels, though an approach to consonantal sounds is frequently manifested.

Early man, according to the conclusions of philological science, possessed a language of the kind here described, consisting of a few calls and cries, each conveying some general information or indicating some emotion. As man's needs increased, the number of these vocal utterances increased correspondingly, with a growing variety of consonantal sounds. In time, it is probable that a considerable vocabulary thus came into existence, though language still continued but little developed beyond the root-stage of speech.

No human tribe is now in this archaic stage of language; even the lowest savages have progressed beyond it. Yet that it once everywhere existed, is believed to be fully proved by the analysis of existing languages, in each of which a vocabulary of roots emerges as the foundation of all subsequent development. And that this method of speech continued until a somewhat late period in human history seems indicated by one significant fact; this is, that the two most ancient of civilizations—the Chinese and the Egyptian — still possess languages which are but a step beyond the root-stage. The indications are that these peoples rapidly developed from barbarism into civilization at an era when human speech was yet mainly in its archaic stage, and were forced at once to adapt this imperfect instrument to the demands of civilized life, without being able to wait for its natural evolution.

The language of China is strictly monosyllabic, and its words have the generalized force of roots. Yet these vague words have been adapted to the expression of definite ideas in a very interesting manner, which we may briefly consider. The natural development of language consists in expedients for the limitation of the meaning of words, vague conceptions being succeeded by precise and localized ones. This is ordinarily accomplished by the formation of compound words, in which each element limits the meaning of the others. Such an expedient has been adopted in every language except the Chinese and its related dialects. Why it was not adopted by them, is an interesting question, of which a possible solution may be offered.

The study of Chinese indicates that its original vocabulary was a very limited one. The language seems to possess but about five hundred original words. But each of these has several distinct meanings. The ancestors of the Chinese people would appear to have made each of their root-words perform a wide range of duties, instead of devising new words for new thoughts. To advance beyond this primitive stage either an extension of the vocabulary or some less simple expedient was necessary. The Chinese adopted a peculiar method for this purpose, the character of which can be best shown by an illustration. We may instance the word *tao*, which has the several meanings, "to reach," "to cover," "to ravish," "to lead," "banner," "corn," "way," etc. These are modernized meanings. Originally the significance of words was much more vague. At present, however, the word *tao*, if used alone, has the meanings above given; and some method is requisite to show what particular one of them is intended. The difficulty thence arising is partly overcome by the device of tones, of which eight are occasionally, and four are commonly used. The tone in which a word is spoken — whether the rising, the falling, the even, or some other inflection — indicates its particular meaning; and in this way the five hundred original words are increased to over fifteen hundred.

A more important device is that of combination. Two words having some similarity or analogy in one of their meanings are joined, and a special meaning is thus indicated. Thus the word *tao*, above given, has "way" for one of its meanings. *Lu*, out of its eight or ten meanings, has also one signifying "way" or "path;" therefore *tao-lu* means "way" or "road" only. So *ting*, having "to hear" for one of its several meanings, is confined to this meaning by the addition of *keen*, "to see" or "perceive." General meanings are also gained by the same method. Thus *fu*, "father," combined with *mu*, "mother," yields *fu-mu*, "parents." *Khing*, "light," with *sung*, "heavy," yields *khing-sung*, "weight." Gender and some other grammatical expedients may be indicated by the same device.

By a consideration of the above facts we can understand why grammatical inflection was never adopted in the

Chinese. Inflection has its origin in word-compounding. But the fathers of the Chinese people seem to have exhausted the powers of word-compounding as a method of increasing their vocabulary. Instead of coining new words to express new things, they seem to have spread their old words over new things, and then limited their meaning by compounding. This gave rise to two important results. It was necessary to retain the integrity of form and meaning of the old monosyllables, since each of them formed a definite part of so many compound words; and it became impossible to express all the intricacy of grammatical relations by word-compounding, since this would have led to inextricable confusion. In consequence, the expedient of the syntactical arrangement of words to express grammatical variations was adopted, and the peculiar Chinese method of speech came into existence.

A Chinese word standing alone has no grammatical limitation. It may be noun, verb, adjective, or adverb at pleasure. Its sense is as indefinite as that of the English word "love," which may be used at will as verb, noun, or adjective. This generalism of sense, found in some English words, is common in Chinese words. The special meaning which each word is intended to convey depends upon its position in the sentence. Every change in its relation to the other words of the sentence gives it a new sense or grammatical meaning. Chinese grammar, therefore, is all syntax. There is no rhetorical freedom in the arrangement of words into sentences. They must be placed according to fixed rules, since any variation in their position gives a new meaning to the sentence. And not only the parts of speech, but the number, gender, and case of nouns, and the mood and tense of verbs, are indicated by the position of the words in the sentence, aided by the use of certain rules of composition and of some defining particles.

The Chinese expedient has been adopted by no other family of language, though the Egyptian vocabulary is almost as monosyllabic and primitive in character. Everywhere else the vocabulary seems to have been extended by coinage of new words, and the principle of word-compounding applied to other uses. The most archaic form of the other types of language is that known as the Incorporating, or Polysynthetic, in use by the American tribes and the Basques of Spain. This is a highly primitive method, and was probably at one time widely spread over Europe and Northern Africa, until replaced by more developed methods of speech.

In the typical incorporating method there are no words, there are sentences only. The verb swallows up both subject and object, with all their modifications. A Basque speaker cannot say "I give." He must say "I give it," in the one word. There is a poverty of the imagination indicated. A hint never suffices; no lacunæ are left for the mind of the listener to fill up. Where we say "John killed the snake," the Basque must say "John, the snake, he killed it;" and all this is welded together into a single complex word. This method is carried to a great extreme in some of the American dialects. The verb absorbs not only the subject, as in Aryan speech, but all the objects, direct and indirect, the signs of time, place, manner, and degree, and all the modifying elements of speech, the whole being massed into a single utterance.

There is little sense of abstract thought in American speech. Everything must be expressed to its utmost details. As an instance we may quote the longest word in Eliot's Indian Bible: *wut-ap-pe-sit-tuk-qus-sun-noo-weht-unk-quoh.* In English we should express this by "kneeling down to him." But in its literal meaning we have, "he came to a state of rest upon the bended knees, doing reverence unto him." Whitney quotes, as a remarkable instance of extension, the Cherokee word *wi-ni-taw-ti-ge-gi-na-li-skaw-lung-ta-naw-ne-li-ti-se-sti*, "they will by that time have nearly finished granting (favors) from a distance to thee and me."

The inordinate length to which words thus tend to grow is somewhat reduced by an expedient of contraction. In forming the compound word the whole of the particle is not used, but only its significant portion. Thus the Algonkin' word-sentence *nadholineen*, "bring us the canoe," is made up of *naten*, "to bring;" *amochol*, "canoe;" *i*, a euphonic letter; and *neen*, "to us."

Savage tribes generally display an inability to think abstractly or to form abstract words, their languages in this respect agreeing with the American. A Society Islander, for instance, can say "dog's tail," "sheep's tail," etc., but he cannot say "tail." He cannot abstract the idea from its immediate relations. A Malay has no separate word for "striking," yet he has no less than twenty words to express striking with various objects, as with thin or thick wood, with the palm, the fist, a club, a sharp edge, etc. This incapacity to express abstract relations is strongly indicated in the American languages, and indicates that they diverged into their special type at a very low level of human speech. The Cherokee, for instance, can use thirteen different verbs for various kinds of washing, but he has no word for the simple idea of washing. He can say *kutuwo*, "I wash myself;" *takungkala*, "I wash my clothes;" *takuteja*, "I wash dishes;" but is quite unable to say "I wash."

All this indicates a very primitive stage of language, in which every expression had its immediate and local application, and each utterance told its whole story. There was no division of thought into separate parts. In the advance of thought men got from the idea "dog" to that of "dog's tail," and from that to "dog's tail wags." They could not think of an action by itself, but could think of some object in action. No doubt all language pursued this course of development up to a certain level. Beyond that point some families of speech began a process of abstraction, gradually dividing thought into its constituent elements. The American type failed to do so, but continued to add modifying elements to its verbal ideas as the powers of thought widened, until language became a series of complex polysyllables. This is the theory advanced by Sayce. All has continued in the original synthetic plan. The secondary method of analysis has not yet acted upon American thought.

Yet it is rather the method of language than of thought that has remained persistent with the Americans. They are undoubtedly able to think more analytically than they speak. The force of their linguistic system has held them to a method of speech which their minds have grown beyond. Every tendency of their language to break up into its elements has been checked by an incorporative compounding, of which traces are yet visible. In two American languages, the Eskimo and the Aztec, the lowest and one of the highest in civilized development, isolation of word-elements has taken place. In these languages a sentence may consist of several words, instead of being compressed into a single word. A process of abstraction exists in the Aztec. Thus the word *ome*, "two," combined with *yolli*, "heart," yields the abstract verb *ome-yolloa*, "to doubt." Through methods such as this the powers of the American type have become increased; yet in character it directly preserves a highly primitive condition of human speech.

The third type of language which we need to consider is that known as the Agglutinative. It is the method used by the Mongolian peoples of Europe and Asia, with the exception of the Chinese and Indo-Chinese, by the Dravidians of India, and, in a modified form, by the Malayans of the Pacific islands.

Agglutination means simply word-compounding for grammatical purposes, without inflectional change of form. In this linguistic method, as in the isolating, the separate words retain their forms intact, but many of them have lost their independence of meaning and become simply modifying particles. To the root-words the others are added as suffixes, with a grammatical significance. The syntax of the Chinese system is here replaced by grammar, the principle of word-compounding having gained a new purpose or significance. In some of these languages each verbal root may be made to express an extraordinary variety of shades of meaning by the aid of suffixes. In the Turkish each root yields about fifty derived forms. Thus if we take the root *sev*, which has the general meaning of "loving," we may obtain such compounds as *sev-mek*, "to love;" *sev-me-mek*, "not to love;" *sev-dir-mek*, "to cause to love;" *sev-in-mek*, "to love one's self;" and so on. By a continued addition of suffixes we arrive at such a cumbrous compound as *sev-ish-dir-il-e-me-mek*, "not to be capable of being made to love one another." Tenses and moods are indicated in the same manner. And there is a second, indirect conjugation, based on the union of the several particles with the auxiliary "to be." In this manner many minute shades of meaning can be expressed. Yet all agglutinative languages are not equally capable in this respect. Thus the Manchu is nearly as bare as the Chinese, while the Finnish and the Dravidian are exceedingly rich. In these languages there is no inflectional variation: every word rigidly preserves its integrity of form. Nor do the particles become welded to the root, and lose their separate individuality, as in Aryan speech. Each seems to exist as a distinct integer in the mind. The only change of form admissible is a euphonic one, in which the vowels of the suffixes vary to conform to those of the root. Thus "to love," is *sev-mek;* "to write," is *yaz-mak*, — *mek* becoming *mak* in harmony with the variation in the root-vowel. This change of vowel is destitute of inflectional significance.

We have yet to deal with the final series of languages, — those organized on what is known as the inflectional method, in which language has attained its highest development and is employed by the most advanced of human races. Here, however, we have two types of language to consider, — those known as the Aryan, and the Semitic: the first, the method employed by the Xanthochroic division of the Caucasians; the second, that in use by the Arabs and other Semites of southwestern Asia.

It is of interest in this connection to perceive how greatly the Aryan languages have prevailed over those spoken by Melanochroic man, despite the probable great excess in numbers of the latter. Of distinctive Melanochroic tongues, the only ones now in existence are the Basque dialect of Spain, and the languages of the Semites and Egyptians, the only Melanochroic peoples who escaped conquest by and assimilation with the Xanthochroi.

It is assumed by many philologists, and not denied by others, that the Aryan and Semitic types of language are Inflectional in the same general sense, and that they may have been derived from one original method of speech, from which they have since developed in unlike directions. Yet the differences between these two types of speech are so radical, and the character of their inflectionalism so essentially different, that it seems far more probable that they have been separate since their origin, and represent two totally distinct lines of development from the root-speech of primitive man.

The common characteristic of Semitic and Aryan speech is their power of verbal variation. There is no tendency to preserve the integrity of form of their words, as in other linguistic types. The root readily varies; and this variation is not euphonic, but indicates a change of meaning. Similar variations take place in the suffixes, particularly in Aryan speech; and the word-compound is welded into a single persistent word, whose elements cease to remain distinct in thought. But aside from this common principle of inflection, the Semitic and Aryan languages differ widely in character, and display no other signs of relationship.

This is what naturally might have been expected if the Melanochroic and Xanthochroic types of mankind were the offspring of different original races, and only mingled after their methods of speech had become well developed.

The steps of progress of Semitic speech have not been traced, and this linguistic method as yet yields little or no evidence concerning the origin of the Melanochroi. The line of development of Aryan speech is more evident. In its most archaic form it is but a step removed from the agglutinative Mongolian type of language, and the latter could readily be changed into an inflectional type closely resembling the Aryan by a single step forward in development. This fact is in close accordance with the inference drawn in our first chapter,—that the Xanthochroi are an outgrowth from the Mongolian race. In some of the agglutinative tongues the principle of word-synthesis is carried to an extreme only surpassed in the American dialects, and compounds of ponderous length are produced. The most archaic forms of Aryan speech greatly resemble these in the extent to which synthesis is carried, and only differ in that their root-forms have become flexible, and that thus a new method of variation of meaning has been introduced, and one which adds the important principle of verbal analysis to the original one of synthesis. Thus in language, as in other particulars, the Xanthochroic Aryans seem a direct derivative from the Mongolian race.

If now we come to Semitic speech, we meet with a type of language which displays no affinity to Mongolian or Aryan speech, and indicates a distinct origin and line of development. The suffixes and affixes which form such essential elements of the Aryan languages are almost unknown to the Semitic. They are used, indeed, but only to a slight extent and as a secondary expedient. The method of word-compounding, which is so widely used in all the languages we have so far considered, is almost absent from the Semitic type, which in this respect fails to come up to the level even of the Chinese. The ruling principle in Semitic speech is inflectionalism pure and simple. It is characterized by an internal or vowel inflection of the root, which has proved so valuable an expedient as greatly to reduce the necessity of word-compounding, and render the use of suffixes and affixes unimportant. The distinction between Aryan and Semitic inflection becomes thus clearly outlined. The former possesses vowel-inflection of the root to a slight degree. Yet this seems principally of modern origin, while the use of the suffix is the ruling grammatical expedient. On the contrary, in Semitic speech vowel-inflection rules supreme, and word-compounding is so little used that it perhaps formed no part of the original linguistic idea, but is of later introduction.

To so great an extent do the vowels of the Semitic root change, and so persistent are the consonants, that the latter are considered as the actual root, there being no basic root-forms with persistent vowel or vowels. A Semitic root thus usually consists of three consonants, and changes its significance with every variation in the vocalization of these consonants. There is some reason to believe that originally the roots contained two consonants only; but at present the three consonants are almost invariably present.

As an illustration we may offer the frequently quoted Arabic root *q - t - l*, which has the general sense of "killing." The signification of this root is variously limited by the vowels used. Thus *qatala* signifies "he kills;" *qutila*, "he was killed;" *qutilu*, "they were killed;" *uqtal*, "to kill;" *qatil*, "killing;" *iqtal*, "to cause killing;" *quatl*, "murder;" *qitl*, "enemy;" *qutl*, "murderous;" and so on through numerous other variations. It may readily be seen how essentially this linguistic method differs from the Mongolian and the Aryan, with their intricate use of suffixes. In the Semitic not only special modifications of sense, but the grammatical distinctions of tense, number, person, gender, etc., are indicated in the same manner. The system is extended to cover almost every demand of language. Each Arabic verb has theoretically fifteen conjugations, of which ten or twelve, each with its passive form, are in somewhat common use. Suffixes, prefixes, and even infixes are moderately employed, but Semitic words never add ending to ending to the formation of long and intricate compounds, as in Aryan and Mongolian speech.

The Semitic languages, comprising the Hebrew and Arabic, the ancient Assyrian, Phœnician, etc., are remarkable for their rigidity. For centuries they persist with scarcely a change. This seems, indeed, a necessary consequence of their character. The root is the most unchanging of verbal forms, and the root is the visible skeleton of every Semitic word. Hardly a single compound Semitic word exists, while variation of form takes place with exceeding slowness.

The Semitic type of language thus points to the speech of primitive man as directly as does the Chinese. It is root-language to a very marked extent, and does not occupy the high position in linguistic development which is often ascribed to it. Its superiority to the Chinese consists in the adoption of a superior expedient, — that of root-inflection, which served all linguistic purposes, and checked further development by rendering unnecessary the employment of other expedients, as in the remaining types of speech.

It has consequently retained its archaic method with rigid persistency.

The Melanochroic people of Africa possess what is usually considered a distinct type of language, known as the Hamitic, and spoken by the ancient Egyptians, the modern Copts, and by the Berbers of the Sahara region from Egypt to the Atlantic. These languages are related to the Semitic family. Many of their roots are similar to Semitic roots, and in grammatical structure there are marked traces of Semitic affinity. Yet there are characteristics differing from the Semitic. It may be that the two types of speech were derived from a single source and have developed somewhat differently. The Egyptian language is monosyllabic, and its forms are almost as rigid and archaic in structure as those of the Chinese. This monosyllabilism has been traced by some writers to a Nigritian source. The monosyllabic character pertains to several of the Negro languages; and the fact that their vocabularies differ from the Egyptian proves nothing, since savage vocabularies often change with great rapidity.

This suggestion is in accordance with the idea advanced in regard to the origin of the Melanochroic race. In fact, our consideration of the languages of mankind leads to some interesting conclusions. The two primitive races, the Mongolian and the Negro, probably both used originally a root-method of speech. Each of them, according to our view of the case, developed into a very ancient civilization, — the Chinese and the Egyptian. These civilizations came into existence ere language had advanced far beyond its archaic root-condition; and in the adaptation of this imperfect method of speech to the needs of man in his earliest civilized stage, roots continued the main constituent of language, and were variously dealt with to express the multitude of new ideas that arose. The root-language from which came that of Egypt may have, in another region, developed the highly effective system of root-inflection of Semitic speech. Alike in the Semitic and the Hamitic linguistic types, the use of suffixes and affixes prevails to a limited extent; and in this respect they are in harmony with the Nigritian languages, — their possible ancestral stock, — in which the agglutinative principle has attained some slight development. But the separation of these several types must have taken place at a very remote date, while language was yet but little developed beyond its archaic stage.

In the Mongolian languages root-inflection failed to appear, and the principle of word-compounding took its place as the ordinary expedient. We have traced this line of development of language through its arrested stage in Chinese, and its unfoldment in American and Mongolian speech, to its culmination in Aryan,— a linguistic type which seems to be in direct continuity with the Mongolian agglutinative method. This consideration leads to the same conclusion which we reached in studying the races of mankind. We seem to perceive two original races, the Mongolian and the Negroid, each with its archaic type of speech, closely resembling each other originally, but pursuing different lines of development, the former reaching its final stage in the speech of Xanthochroic man,— the highest outcome of the Mongolian race; the latter in the speech of the Semites, — the highest outcome of the Negroid race. It remains, in conclusion of this chapter, to consider the development of the Aryan type of speech, — the most effective instrument of intellectual expression yet attained by man.

In the Aryan languages alone has verbal analysis become a prominent characteristic. In the Semitic tongues there is no analysis, and almost no synthesis. The same may be said of the Chinese and its cognate dialects. In the other languages of Asia, and those of Europe and America, synthesis is a prevailing characteristic, it reaching its culmination in the interminable American compounds. It is less declared in the Mongolian tongues, but in none of them does word-analysis appear. This is only found as an active principle in the Aryan of all the families of speech. In the Aryan languages it has always been a ruling characteristic, though it is not strongly declared in the most archaic of these dialects. No tendency to preserve the integrity of form in words exists, and abrasion has gone steadily on, reducing the length of verbal elements, and wearing down or breaking up compound words into monosyllables, until some Aryan tongues have gained a monosyllabilism approaching that of the Chinese. It is this analytic tendency which has produced and constitutes the Aryan method of inflection, and in which it is strongly contrasted with the vowel-inflection of Semitic speech.

From its origin, the Aryan type of speech has manifested the double power to build up and to break down, and these powers have been continually in exercise. It is an interesting fact, however, that the building-up or word-combining tendency long continued the more active, and yielded such highly complex inflectional languages as the Sanscrit and the Greek. The variation from the Mongolian method was not yet decided, and the synthetic principle continued in the ascendency. But throughout the succeeding period, down to the present time, the abrading or analytic tendency has been the more active, and languages of very simple structure have arisen. This is most strikingly the case in English speech, but it is also strongly declared in the Latin derivative languages, in modern Persian and Hindu, and to some extent in modern Greek and German. It appears to have met with most resistance in Slavonic speech, in which the synthetic tendency has vigorously retained its ascendency.

In all the ancient Aryan tongues the use of word-combination for grammatical expression was vitally active. Highly complex languages arose, which are often spoken of with an admiration as if they had attained the perfection of linguistic structure, and as if modern languages were barbarous in comparison. And yet they are superior to agglutinative speech only in the fact that they permit verbal variation. They are cumbersome and unwieldy to modern tongues, which have become fitted to the use of a simpler and swifter speech.

No sooner did the vigor of word-combination grow inactive, checked probably by the complexity it had evolved, than the analytic tendency became prominent, and began to break down the cumbrous compound words into their elements. The pronoun was separated from the verb. Particles were torn off and used separately. Auxiliaries came into more frequent use. Analysis rose into active competition with synthesis. Yet this did not proceed rapidly in the ancient historic period. That was an age of literary cultivation, in which language became controlled by standards of authority, and its variation was greatly checked. The most active analytic change was that displayed by the Latin, the speech of a highly practical people, who were more attracted to ease and convenience of utterance than to philosophic perfection of grammatical method.

As the synthetic principle had originated during the primal period of Aryan barbarism, and reached its highest development during the ancient era of literary cultivation, so a second period of barbarism seemed essential to any rapid action of the analytic principle. This period came. The ancient civilizations vanished, and a long-continued era of mental gloom overspread the Aryan world. Throughout this Middle-Age period the restraining influence of literature ceased to act. Nearly all the literary cultivation that remained was restricted to the classical Latin and Greek in the West, and Sanscrit in the East. Every check to dialectical change was removed, and language varied with the utmost activity.

This variation, in Europe, was greatly aided by the forcible mingling of peoples speaking unlike dialects. In France, Italy, and Spain the Latin became exposed to the influence of barbarian invaders accustomed to a different speech. The complex words, with their intricate significance, proved a burden to these new speakers; they became broken up into their elements.[1] When, at a later period, the minds of men became again cultivated, and thought regained some of its vanished powers, the analytic tendency held its own; the old synthetic process had lost its force. Auxiliaries and words of relation came more and

[1] Philologists believe that a barbarous Latin, analogous to the jargons known as Pigeon English and Lingua Franca, became the medium of communication between the conquerors and their subjects, the grammatical perfection of the classic Latin disappearing, and being replaced by a linguistic method of great simplicity. Similar conditions may have attended the mingling of dissimilar languages in England, Persia, and elsewhere; yet such an influence could but have accelerated what seems the natural tendency of the Aryan type of language toward analytic methods of speech, since this has shown itself in places and periods in which no such specially favoring influence existed.

more into use. Complex ideas, instead of being condensed into single words, as of old, were expressed by groups of words, each of which constituted a separate element of the idea. A distinct and highly valuable step forward in the evolution of language had been gained. As in ancient writing the characters at first expressed ideas, then words and syllables, and finally alphabetic sounds, so thought became divided into its prime elements, and instead of spoken words expressing complete ideas, as in American speech, or sectional parts of ideas, as in agglutinative and early inflectional speech, they became reduced into the component elements of ideas. A sort of chemical analysis of thought had taken place. Thought had, if we may so express it, been reduced to its alphabetic form.

This, the highest, and probably the final, stage in the evolution of language, has nowhere gained its complete development. In some languages, as in the modern German, which remained unaffected by transplantation and mixture with a foreign tongue, the synthetic principle is still vigorously active. The analytic has gained its fullest development in modern English. This tendency, indeed, was strongly at work upon the Anglo-Saxon long before its intermixture with foreign elements. Of all Aryan dialects it showed the most active native inclination to analysis. The reduction of words to monosyllables, the loss of inflectional expedients, and the use of separate auxiliaries, pronouns, prepositions, etc., made considerable progress in the long dark period before the Norman Conquest. This latter event intensified the change of method. The forced mingling of two modes of speech, each already tending to analysis, and each with but little literary cultivation, could not but have an important effect.

The synthetic forms rapidly decreased, and there finally issued a language of elementary structure, largely monosyllabic, almost devoid of inflection, and to some extent displaying a reversion to the root-stage of human speech.

Such is the English of to-day, — the most complete outcome of linguistic analysis yet reached, the highest stage attained in the long pathway of verbal evolution. At first glance it seems to have moved backward instead of forward. It has approached the Chinese in its loss of inflection, its monosyllabilism, and its partial replacement of the grammatical by the syntactical arrangement of the sentence. Yet this is no real reversion. Our pride in the richness of Aryan speech as compared with the poverty and imperfection of the Chinese is apt to blind us to the fact that the Chinese system has features of decided value. Similar features have been gained by English speech, while none of the actual advantages of inflection have been lost. In the English we perceive a decided advance toward that simplicity of conditions which marks all highest results. Nearly every inflectional expedient which could be spared, or be replaced by an analytic expedient, has been cast off. The inflection of nouns has almost vanished. That of adjectives has quite disappeared. Only in the pronouns does inflection partly hold its own. The inflectional conjugation of verbs is reduced to a mere shadow of its former self. The utterly useless gender-distinctions which yet encumber the languages of Continental Europe have absolutely vanished.

Nearly all these incubi of language have been got rid of in English, which has moved out of the shadow of the past more fully than any other living tongue. It has in great measure discarded what was valueless, and kept what was valuable in inflectional speech, adopting an analytic expedient wherever available, though freely using the principle of synthetic combination of words where the latter yielded any advantage. It stands in the forefront of linguistic development, possessed of the best of the old and the new, having certain links of affinity with every cultivated type of language that exists, rid of all useless and cumbersome forms, yet possessed of a flexibility, a mingled softness and vigor of tone, a richness of vocabulary, and a power of expressing delicate shades of thought, in which it is surpassed by none, and equalled by few of existing languages.

With a brief comparison of the different Aryan languages this chapter may close. Of all these the Sanscrit of the Vedas is regarded as the most primitive form, the one nearest the original Aryan, as the Vedas themselves are the most ancient record of Aryan thought. It has preserved many archaic forms which are lost elsewhere, and without its aid our knowledge of the ancient conditions of Aryan life would be much reduced. Its syntax is comparatively simple, the dominant ancient method of word-composition taking its place. Its grammatical forms are very full and complete; yet in the modern Hindu dialects the usual reversal of this condition appears. These dialects are marked by an active analytical tendency.

The language of the Zend Avesta of the Persians has strong marks of affinity to the Vedic dialect. In some respects it is more archaic; yet as a whole it is younger in form, the Avestas being of more recent production than the Rig Veda. In modern Persian, however, the analytic tendency is very strongly declared, — more so, perhaps, than in any language except the English, which it resembles in the simplicity of its grammar. It has even gone so far as to lose all distinction of gender in the personal pronoun of the third person. Yet it is said to be a melodious and forcible language. Its great degree of analytic change is probably due to the extensive mixture of races that has taken place on Persian soil.

In regard to the European languages, many efforts have been made to class them into sub-groups. Thus one author ranks the Greek, another the German, another the Slavonic, as nearest the Indo-Persian. One brings the Celtic nearer than the Greek to the Latin, while the more common opinion makes it wholly independent. Of these schemes nothing more need be said, since nothing satisfactory has yet come of them. The Celtic dialects have certain peculiarities not shared by other members of the Aryan family, and are ordinarily looked upon as the most aberrant group. The grammar, indeed, displays features which seem to indicate a non-Aryan influence. The incorporation of the pronoun between the verb and its prefixes in Irish speech has been imputed by Professor Rhys to a Basque influence. Some other peculiarities exist which tend to indicate that the aborigines with whom the Celts mingled exercised a degree of influence upon their method of speech.

Of the Teutonic division, the most striking peculiarity is the possession of the strong, or vowel conjugation, such as we have, for instance, in the grammatical variations of form in "sing," "sang," and "sung." In this respect the Teutonic makes an approach to the Semitic method of inflection, though the principle with it is probably of recent origin. Of the Letto-Slavic group, the Lithuanian is marked by a highly archaic structure. In some few points its grammar is of older type than even the Sanscrit. The Slavonic dialects are characterized by phonetic and grammatical complexity and a great power of forming agglutinative compounds. The indication of language is that the Slavonians have been the least exposed to foreign influence, and are the nearest to the primitive Aryans and to their probable Mongolian ancestors, of any section of the race. As an instance, Sayce[1] quotes from the Russian the two words *Bez boga*, "without God." These can be fused into one word, from which, by the aid of suffixes, we obtain *bezbozhnui*, "godless;" from this is gained the noun *bezbozhnik*, "an atheist," then the verb *bezbozhnichut*, "to be an atheist;" with a host of derivatives, of which may be named *bezbozhnichestvo*, "the condition of being an atheist," and *bezbozhnichestvovat*, "to be in the condition of being an atheist." Certainly the Russian has lost none of the ancient richness of the synthetic method, or descended into what classicists regard as the base abyss of analytic speech. The Finns, with whom the Russians are so mingled in blood, could hardly present an instance of synthesis more complex than the last named. This is precisely the condition we should expect to find in the home-staying section of the Aryan race.

It is to the ancient Greek that we must look for the most logical and attractive unfoldment of the inflectional method. Though eminently capable of forming compounds, it is free from the extravagance displayed by the Sanscrit in this direction, while its syntax has reached a high level of development. Finally, in the Latin, as already remarked, the analytical grammatical tendency is indicated in a stronger degree than in any other ancient Aryan

[1] Introduction to the Science of Language, ii. 95.

tongue. This has been carried forward through the line of its descendants, the Romance languages of southwestern Europe, and is particularly displayed in the French, in which the spoken has run far beyond the written language in its tendency to verbal abrasion. As regards grammatical analysis, however, the English, as already remarked, has gone farther than any modern language, and is only less bare of inflectional forms than its very remote cousin, the Chinese. And it may be said, in conclusion, that the English, while the most advanced in development, has become the most widespread of Aryan languages; it is spoken by large populations in every quarter of the earth; and if any modern language is to be the basis of the future speech of mankind, the English seems the most probable, both from its character and its extension, to attain that high honor.

IX.

THE AGE OF PHILOSOPHY.

THE assertion that the Aryans are intellectually superior to the other races of mankind may be held as not proved by what we have yet related concerning them. In the growth of the primitive conditions of religion, statecraft, industry, language, etc., there was no individual action. These were all results of involuntary evolution, not of purposive activity of the intellect. The democratic character of the Aryan political system, for instance, naturally arose from a primitive stage very closely resembling that attained by the American Indians. The subsequent spirit of liberty of the Aryans seems largely due to the fact that there had also developed among them a democratic or individual religious system, and that, in consequence, there existed no strongly organized and influential priesthood, as elsewhere, to hold their souls in captivity. Their village community system was a natural result of the fact that they became agricultural ere any progress in political organization had been made. The same result arose from the same conditions in America. In the primitive agricultural civilizations of Egypt and China, on the contrary, the political organization probably preceded the development of agriculture, and patriarchism became established. The same thought applies to the Aryan language. Its superiority may be due to the fact that out of the several possible methods of speech-evolution the Aryans chanced to adopt the one most capable of high development, and which has, in consequence, continued to unfold its capabilities while the other types have long since reached a stage of rigid specialization.

And yet all this must be more than the effect of mere chance. It would be very surprising if a single race should have blundered into the best methods of human development in all directions. Though in regard to the matters so far considered there is no probability that individuals exercised any important voluntary control over the development of institutions, yet the collective intellect of the Aryans could not have been without its directive force. It undoubtedly served as a rudder to guide the onward progress of the race and prevent this from becoming the mere blind drift of chance. This much we clearly perceive, — that the Aryans nowhere entered into a rigidly specialized state. In all the unfoldment of their institutions they pursued that mid line of progress which alone permits continued development. If we compare the only one of the non-Aryan civilizations that has survived to our time, the Chinese, with those of Aryan origin, this fact will become evident. In all respects, in language, politics, religion, etc., the Chinese early attained a condition of strict specialization, and their progress came to an end. For several thousand years they have remained stagnant, except in the single direction of industrial development, in which some slow progress has been made. But in all these respects the Aryans have continued unspecialized, and their development has been steadily progressive. This progress yet actively continues; while there is no hope for China, except in a complete disruption of its antique system and a deep infusion of Aryan ideas into the Chinese intellect. This general Aryan superiority is indicative of a highly active and capable intellect, even though no one mind exercised a controlling influence. The general mentality of the race, the gross sum of Aryan thought and judgment, must have guided the course of Aryan evolution and kept our forefathers from those side-pits of stagnation into which all their competitors fell. During its primitive era the Aryan race moved steadily forward unto a well-devised system of organization which formed the basis of the great development of modern times.

It is our purpose now, however, to consider the unfoldment of the intellect at a higher stage, — that in which individuality came strongly into play, single men emerged from the mass of men, and great minds brought their strength to bear upon the movement of human events. It is here that the superiority of the Aryan intellect makes itself first specially apparent. The mentality of the race developed with remarkable rapidity, and yielded a series of lofty conceptions far beyond the products of any other race of mankind. A brief comparison of the attainments of the ancient Aryan intellect with the mental work of contemporary nations cannot fail to show this clearly. We shall here concern ourselves with the philosophical productions of the race, before considering their more general literary labors.

As already said, the human intellect is primarily made up of two great divisions, the reason and the imagination, which underlie its more special characteristics. Reason is based on the practical, imagination on the emotional, side of thought. These are the conditions which we find in a specially developed state in the two most distinguishable primary races of man, the Mongolian and the Negro. The Mongolian is practical man, the Negro emotional man. In each of these two races the quality named is present in a marked degree, while the other quality has attained only a minor development. The same rule applies to the two race-divisions of the Caucasians, considered as derivatives respectively of the two original races. The pure Xanthochroi strongly display the Mongolian practicality; the pure Melanochroi the Negro emotional excitability. Yet the one has unfolded into reason, the other into imagination. But for the complete development of these high faculties a mingling of the two sub-races seemed requisite. The practical mental turn of the Xanthochroi needed to be roused and invigorated by an infusion of the excitable fancy of the South; the fanciful mentality of the Melanochroi to be subdued and sobered by an infusion of the practical judgment of the North. As a result arose the mingled reason and imagination of the Aryan intellect, each controlling, yet each invigorating the other, until through their union mentality has reached the acme of its powers, and human thought has made the whole universe its field of activity.

Of the non-Aryan civilizations which have attempted to enter the field of philosophy, three only need be named, — the Chinese, the Egyptian, and the Babylonian. As for the American civilizations, they were when destroyed still in the stage of mythology. Everywhere, indeed, mythology appears as the result of the earliest effort of the human mind to explain the mysteries of the universe. The forces and forms of Nature are looked upon as supernatural beings, with personal histories and man-like consciousness and thought. This is but little displayed by the practical

Chinese, who had not imagination enough to devise a mythology. We find it much more strongly manifested by the Egyptians, who had much of the fervor of the Melanochroic fancy.

It was with the detached and often discordant mythologic figments, produced through a long era of god-making, that philosophy first concerned itself. When men had passed through the ancient era of blind worship of the elements, and begun to think about the theory of the universe which had grown up involuntarily during the long preceding centuries, they were not slow to perceive its incongruity. Everywhere gods crowded upon gods. Their duties and attributes clashed and mingled. Their names flowed together. Their histories overlapped each other. All was utter confusion and discord of ideas. It was very apparent that there must be error somewhere. Heaven and earth could not be governed in this chaotic fashion. Some order must exist beneath this interminable show of disorder.

It is not difficult to understand how this confused intricacy had arisen. There is reason to believe that in ancient Arya, though many gods were recognized, each worshipper addressed himself to but one deity at a time, whom he looked upon as supreme, and whom he invested with all the deific attributes. This system, named "henotheism" by Max Müller, is the one we find in the hymns of the Rig Veda. In succession the different gods of the Aryan pantheon are supreme deities to these antique singers.[1]

[1] "It would be easy to find, in the numerous hymns of the Vedas. passages in which almost every single god is represented as supreme and absolute. Agni is called 'Ruler of the Universe.' Indra is celebrated as the strongest god. It is said of Soma that 'he conquers every one.'" — *Max Müller.*

Men's minds seemed not sufficiently expanded actually to grasp the thought of more than one god at a time, though they acknowledged the existence of many. This ascription of the various duties, powers, and attributes of the deity to so many different beings, necessarily produced considerable confusion, which increased with the growth of mythologic fancies. It grew with particular rapidity in Greece, since the actively commercial Hellenes imported new gods from Phœnicia, Assyria, and Egypt, and mingled them with the tenants of the ancient Aryan pantheon, until the confusion of ideas became somewhat ludicrous.

It is interesting to find that in the earliest efforts of men to obtain a philosophical idea of the universe the thinkers were still ardent believers in mythology, and their efforts were limited to an attempt to divide the duties of celestial government among the several deities, and introduce order into the deific court. This stage of thought we find vaguely indicated in Egypt and Babylonia, and more definitely in Greece; but it yielded no important results in any of these regions. The disorder was too great, and the mingling of the deific stories too intricate, to admit of any success in their rearrangement. In Egypt and Greece, indeed, thought soon passed beyond this stage; the gods were left to the unquestioning worship of the people, and thinkers began to devise systems of philosophy outside the lines of the old mythology. The same was the case in India; but nothing that can be called a philosophy of the universe arose among the Semites. Certain highly fanciful cosmological ideas were devised; but the religious system remained largely in the henotheistic stage. Of the superior gods of the old mythology, each Semitic nation selected one as its supreme deity, or perhaps raised to this honor its own divine ancestor after his ancestral significance had become greatly dimmed. These supreme deities became each the Lord, the King, the Ruler. The cloak of myth fell from their mighty limbs, and left them standing in severe and unapproachable majesty, — the sublime rulers of the universe, for whom it would have been sacrilege to invent a history, and to whom there was left nothing of human frailty, and little of human sympathy. Such was the course of Semitic thought. It devised no philosophy, yet it evolved, as its loftiest product, a strict monotheism, — a conception of the deity that grew the more sublime as it divested itself of imaginative details.

In two branches of the Aryan people the effort to organize mythology and work over this old system of belief into a consistent theory of the universe attained some measure of success. These were the Persians and the Teutons. The Persian system, indeed, which grew up among the followers of Zoroaster, dealt but little with the old mythology, but devised a new one of its own. Yet its philosophy was largely mythological, and it bears a resemblance to the Teutonic so marked as to make it seem as if some of their common ideas were of ancient Aryan origin. These two philosophies of mythology, the only complete ones that have ever been devised, are of sufficient interest to warrant a brief description.

The Persian system is only partly to be ascribed to Zoroaster. Its complete unfoldment is the work of the thinkers of a later period. Several of the steps of its development are yet visible. A comparison of the Avesta with the Vedas shows interesting indications of a religious schism between the Hindu and the Iranian sects. The *Devas*, the "shining ones," of the Hindus became the

Daevas, the "demons," of Iran. On the contrary, the Hindu demons, the *Asuras*, became the *Ahuras*, the gods of the Iranians. One of the Ahuras, a Mazda, or world-maker, was chosen as the special deity of the Zoroastrian faith, which originally had a monotheistic character, — or rather it was in principle dualistic, since Ahura-Mazda comprised two natures, and combined within one personality the double deific attributes of good and evil.

At a later period these attributes unfolded into two distinct beings, and a new supreme god was imagined, — Zarvan Akarana (Boundless Time), the primal, creative power. The mythologic philosophy, as finally completed, was briefly as follows. In the beginning the Absolute Being, Zarvan Akarana, produced two great divine beings, — Ahura Mazda, and Angra Mainyas, or, as ordinarily named, Ormuzd and Ahriman. These were respectively the lords of light and darkness, — Ormuzd a bright, wise, all-bountiful spirit; Ahriman an evil and dark intelligence. From the beginning an antagonism existed between them, which was destined to continue until the end of time. Zarvan Akarana next created the visible world, destined to last twelve thousand years, and to be the seat of a terrible contest between the great deities of light and darkness.

Ormuzd manifested his power by creating the earth and the heavens, the stars and the planets, and the Fravashi, the host of bright spirits; while Ahriman, his equal in creative ability, produced a dark world, in opposition to the world of light, and peopled it with an equal host of evil spirits. This contest between the two great deities was to last until the end of time. Yet the Spirit of Gloom was inferior in wisdom to the Spirit of Light, and all his evil actions finally worked to aid the victory of Ormuzd.

Thus the bull, the original animal, was destroyed by Ahriman; but from its carcass man came into being under the creative command of Ormuzd. This new race increased, while the earth became peopled with animals and plants. Yet for every good creation of Ormuzd, Ahriman created something evil. The wolf was opposed to the dog, noxious to useful plants, etc. Man became tempted by Ahriman in the form of a serpent, and ate the fruit which the tempter brought him. In consequence, he fell from his original high estate, and became mortal and miserable. Yet the human race retained the power of free-will: they could choose between good and evil; and by their choice they could aid one or the other of the great combatants. Each man became a soldier in the war of the deities.

Between heaven and earth stretched a great bridge, Chinvat, over which the souls of the dead must pass. On this narrow path the spirits of the good were conducted by Serosh, the archangel who led the heavenly host. But the evil souls fell from it into the Gulf of Duzahk, to be tormented by the Daevas. Those whose evil deeds had not been extreme might be redeemed thence by prayer; but the deepest sinners must lie in the gulf until the era of the resurrection. At the end of the great contest a terrible catastrophe is to come upon all created things. Man will be converted from his evil ways. Then will follow a general conflagration. The earth will melt with fervent heat, and pour down its molten floods into the realm of Ahriman. A general resurrection of the dead will attend this conflagration. In the older portions of the Avestas this seems to be restricted to the soul; but in the newer portions the resurrection of the body is indicated. The souls are clothed upon by new flesh and bones; friends recognize each other; the just are divided from the unjust; all beings must pass through the stream of fire which is pouring down from the molten earth. To the good it will feel like a bath of warm milk; but the wicked must burn in it three days and nights. Then, purged of their iniquity, they will be received into heaven. Afterward Ahriman and all his angels will be purified in the flames, all evil will be consumed, all darkness banished, and a pure, beautiful, and eternal earth will arise from the fire, the abode of virtue and happiness for evermore.

It is hardly necessary here to call attention to how great an extent the Semitic cosmogony and religious myths are counterparts of this Aryan scheme. It will suffice to say that the Semites seem to have borrowed everything in their creed that approached an effort philosophically to explain the universe. The later Semitic creed, that of Mohammed, is a medley of pre-existing thought. Even the Persian bridge of the dead appears in it as Al Sirat, the razor-edged road from heaven to earth. The Koran is full of extravagant fancies, but devoid of original ideas. It is the outcome of the Arabic type of mind, in which fancy is exceedingly active, but in which the higher powers of the reason seem undeveloped.

In the Teutonic myths are displayed a system of the universe which bears certain striking points of resemblance to that of Persia, though utterly unlike it in its details. The general ideas of these myths, indeed, are common to all the Aryan mythologies, and must have been current in ancient Arya. Thus the Persian Chinvat, or Knivad, the bridge of the dead, is paralleled by the Teutonic Bifrost and the Vedic "path of Yama," the "cows' path," which passes over the abyss of Tartarus to the land of the wise Pitris, the fathers of the nation. In this mythical bridge both the Milky Way and the rainbow are symbolized. Such was the explanation given to these striking natural phenomena by our imaginative and unscientific forefathers.

But with the Teutonic tribes, and particularly with their Scandinavian section, we have to do with a people very different in situation and culture from the Persians. The latter were a partly civilized people, the former fiercely barbarous. The latter dwelt in a temperate region, the former in an arctic land, where ice and cold were the demonic agents of man's torment. Yet the strong Aryan intellect stirred in their minds, and from their ancestral myths they wrought out a coherent system of the universe, — the wildest and weirdest that it ever entered the brain of man to conceive. It was mythology converted into philosophy; but it was the mythology of the barbaric and warlike North, with the breath of the arctic blasts blowing through it, and the untamed fierceness of the Norman vikings in its every strain. This system, as fortunately preserved to us in the Eddas of Iceland, and perhaps mainly of Scandinavian development, may be here briefly given, omitting its many side-details. Everywhere it is full of warfare. The soul of man is free to combat with the powers of Nature. The gods are always at war. Sunshine and growth combat with storm and winter. Frost opposes fire. Light and heat are in endless conflict with darkness and cold. The Jotuns, the ice-giants, are the demons of Scandinavia. The forces of the winter everywhere bear down upon those of the summer, and finally overwhelm and destroy them. But this battle of the elements is wrought into a weird story of the conflict of gods and demons, in which the traces of its origin are nearly lost.

In the beginning there lay to the south the realm of Muspell, the bright and gleaming land, ruled by Surtr of the flaming sword, the swart god. To the north lay Niflheim, the land of frost and darkness. Between them was Ginunga-gap, — a yawning chasm, still as the windless air. From the ice-vapor that rose from Hvergelmir, the venom-flowing spring of Niflheim, and mingled with the spark-filled air of Muspell, was born, in Ginunga-gap, the giant Ymir, the parent of the Jotuns, or frost-giants. But with Ymir came the primal animal to life, — the cow, whose milk nourished the giant. She licked the salt rime clumps, and forth came Buri, a great and beautiful being, the ancestor of the gods. After much gigantic medley the gods slew Ymir, whose blood drowned all his evil race except a single pair, who escaped, to give rise to a new Jotun crew. And now the gods began their creative work. The slain Ymir was flung into the chasm of Ginunga-gap. Here his body formed the earth, his blood the ocean, his bones the mountains, his hair the trees. The sky was made from his arched skull, and adorned with sparks from Muspell. His brain was scattered in the air, and became the storm-clouds. A deep sea was caused to flow around the earth, — the grand, mysterious ocean, the endless marvel to the Northern mind. The escaped giants took up their abode in Jotunheim, the frost-realm of the arctic seas, the ocean's utmost strand. Between Atgard, this outer realm, and Midgard, the habitable earth, the brows of Ymir were stretched as a breastwork against the destructive powers. From earth to heaven extended the rainbow bridge Asbru, the Æsirs' bridge, or Bifrost, the "trembling mile." Every day the gods ride up this bridge to Asgard, the Scandinavian heaven. They ride to the Urdar fount, which flows from beneath the roots of the great ash-tree of life, Yggdrasil, there to take counsel concerning the future from the three maidens — the Past, the Present, and the Future — who daily sit beside the celestial fount.

The first human pair were made by the gods from two trees on the sea-shore; their names were Ask and Embla. To them Odin gave spirit, Hœnir understanding, Lodurr blood and fair complexion. They received Midgard for their abode. From them sprang the human family. But in heaven and earth perpetual warfare raged. The gods and the frost-giants were endlessly at war. But as Ahriman was overcome and fettered by Ormuzd, so Loki, the wolf, the deceiver of the gods, was bound in chains, and a serpent placed above him to drop venom on his face. This venom as it dropped was caught by his wife in a vessel. Only when she went away to empty the vessel did the poison-drops reach his face. Then he writhed in his chains, and earthquakes shook the solid globe.

It is fated that all this shall end in a mighty conflict, in which gods and demons alike shall be slain, and heaven and earth disappear. Ragnarok, the "Twilight of the Gods," shall be ushered in by a winter three years long. The crowing of three mighty cocks shall proclaim the fateful day. Thereat shall the giants rejoice, the great ash take fire, and all the powers of destruction — wolves, sea-monsters, hell hounds, and the like — rush to the dreadful fray. Heimdal, the guardian of the rainbow, shall sound his mighty horn to warn the gods, who shall rush to counsel beneath the tree Yggdrasil, that meanwhile trembles to its deepest roots. From the East shall come the frost-giants in a mighty ship, while another ship, made of dead men's nails and steered by Loki, brings the troop of ghosts. Surtr of the flaming sword, the ruler of Muspell, shall thunder with his swart troop over the bridge of the gods, his fiery tread kindling it into a consuming flame as he rides in grim fury to the stronghold of the deities.

Now meet the combatants, — the gods and the heroes of Valhalla on the one side; on the other the giant crew, led by Fenrir the great wolf, the mighty Midgard serpent, the terrible Loki, and Hela, the goddess of death. Dreadful is the combat. Odin fights with the wolf, Thor with the serpent, Freyr with Surtr, Heimdal with Loki. Death everywhere treads; Odin, the king of the Æsir, is swallowed into the yawning gape of his monstrous antagonist. One by one the mighty combatants fall, while Surtr stalks terribly over the field, spreading everywhere fire and flame. All is consumed, the stars are hurled from the sky, the sun and the moon devoured, and the universe sinks in utter ruin.

Possibly here ended the original myth. It is an ending in consonance with the grim temper of the vikings of the North. But as we have it in the Edda, it goes on to a future state like that of the Persian myth. After the ruin of Ragnarok a new heaven and earth shall rise from the sea. Two gods, Vidar and Vali, and a man and woman shall survive the conflagration and people the new universe. The sons of Thor shall come with their father's hammer and end the war. Balder the beautiful god and the blind god Hödr shall come up from hell, and a new sun, more beautiful than the old, shall gleam in the sky. This is, briefly told, the Scandinavian scheme of the universe, — a rude and fierce one, yet instinct with a vigor of imagination shown nowhere by men of non-Aryan blood. It is the only pure organization of mythology into a coherent system that exists; for the Persian myth includes philosophical ideas which fail to enter into the ruder Scandinavian story of the deeds of the gods, and Greek mythology never fairly emerged from its abyss of confusion.

If now we come to consider the mental evolution of more civilized man, we find everywhere mythology left for the amusement of the vulgar horde, while the enlightened few devise purely philosophical explanations of the mystery of the universe. But in comparing the philosophies of the various civilized nations, the Aryans will be found to soar supremely above the level of all alien peoples. Only two such peoples, Egypt and China, have devised anything that deserves the title of philosophy; for nothing of the kind exists in any of the Semitic creeds. The utmost we find in Babylonia is an effort to form a cosmology of strictly mythologic character, — a highly confused affair as imperfectly given by Berosus. The later attempt made by Mohammed is, so far as it is original, an absurd tissue of extravagant fancies. There is nothing to indicate the least native tendency of the Semitic mind toward philosophy. All their philosophy is borrowed, and has deteriorated in their hands. It was by stripping the idea of deity of all mythologic and philosophic figments, and leaving it in its bare and unapproachable majesty, that the Semitic intellect reached its highest flight, that symbolized in the Jehovah of the later Hebrews.

The Egyptian priesthood, on the contrary, appears to

have devised a somewhat advanced system of philosophy, which bears a singular resemblance to that of Brahmanism, though very far below it in the power and clearness of thought displayed. The transmigration hypothesis, and the theory of emanation and absorption of souls, are both indicated in the Egyptian system, though vaguely, and overlaid with mythological absurdities. There is here none of the clear-cut reasoning of the Hindus, but an uncertain wandering of thought from which it needs considerable ingenuity to extract the idea it conceals. The well-known *Ritual of the Dead* is the source of our knowledge of these confused ideas. A copy of this work, more or less complete, was placed in every Egyptian coffin, while its more important passages were written on the wraps of the corpse and engraved on the coffin. It was necessarily so placed, according to their belief, since it contained the instructions requisite to convey the soul of the deceased safely past the dangers of the lower world. Throughout the whole story physical ideas struggle with metaphysical. The Egyptian mind failed definitely to rise above the level of the world of sense.

After death the soul descends with the setting sun into the nether world. There it is examined and its actions weighed before Osiris and the terrible forty-two judges. If it can declare that it has committed none of the forty-two sins, it is permitted to pass on. It has with it in the Ritual prayers to open the gates of the various lower realms, and to overpower opposing spirits and monsters. It must be able to name everything which it meets, and to recognize the gods it encounters. Here we have indications that the soul is returning to its natal home, and recalling its ante-terrestrial memories. All this the Ritual teaches the spirit, and also provides it with a charm to unlock the gates that lead to the fields of Ra, the sun-god. Finally, if the heart prove not too light, and the soul pure, the members of the body, renewed and purified, are returned to the spirit, and the waters of life are poured upon it by the goddesses of life and the sky. It finally enters the realm of the sun, and vanishes in a highly vague identification with Osiris, or with the deific powers generally. The idea of metempsychosis also confusedly mingles with this, and animal-worship seems at the basis of the Egyptian mythology. The thought of Egypt never fairly rises above the body. There is no entrance into that pure atmosphere of soul-existence in which the Hindu philosophers are at home.

The philosophical system of China is a curious one, which, however, we can but very briefly describe. It had a continuous development, its antique basis being in the mystical symbols of Fu-hi, — a monarch of some such dubious date as 2800 B. C. These symbols consisted simply of a whole and a divided line, constituting the diagram (——, — —). These lines were variously combined, so as to make in all sixty-four combinations. On this strange arrangement of lines, which very probably was connected with some ancient system of divination, an abundance of thought has been exercised, and the whole system of Chinese philosophy gradually erected. The first great name in this development is that of Wan Wang, of about 1150 B. C. Being imprisoned for some political offence, this antique philosopher occupied himself in studying out the meaning of these combinations. The result of his reflections was the *Y-King*, — among the most ancient and certainly the most obscure and incomprehensible of all known books. The Y-King comprises four parts. First are the sixty-four diagrams, each with some name attached to it; as heaven, earth, fire, etc. Second, are a series of obscure sentences attached by Wan Wang to these diagrams. Third, we have other ambiguous texts by Tcheou-king, the son of Wan Wang, the Chinese Solomon. Fourth, are a host of commentaries, many centuries later. The whole forms an intricate system of philosophy, which is based on the idea of the duality of all things. The whole lines represent the strong, the divided lines the weak, or the active as contrasted with the passive. These indicate two great primal principles, — *Yang*, the active, *Yin*, the passive, — which owe their origin to the *Tai-keih*, the first great cause. All existence comes from the Yang and the Yin: heaven, light, sun, male, etc., from the Yang; earth, darkness, moon, female, etc., from the Yin. This development of the idea is mainly the work of the later commentators. Tai-keih, or the grand extreme, is the immaterial producer of all existence. Yang and Yin are the dual expression of this principle, — Yang the agency of expansion, Yin that of contraction. When the expansive activity reaches its limit, contraction and passivity set in. Man results from the utmost development of this pulsating activity and passivity. His nature is perfectly good; but if he is not influenced by it, but by the outer world, his deeds will be evil. The holy man is he with full insight of this twofold operation of the ultimate principle, and of these holy men Confucius was the last. Such is the developed philosophy of the Y-King as expressed by Choo-tsze (1200 A. D.), — one of the latest of the many commentators who have sought to unfold the Fu-hi symbols into a philosophy of the universe.

Of the best-known Chinese philosophers, Confucius and Lao-tsze, the system of the former was simply a creed of morals; that of the latter was but an unfoldment of the dual idea. To Lao-tsze the primal principle was a great something named the *Tao*, concerning which his ideas seem exceedingly obscure. Tao was the unnamable, the empty, but inexhaustible, the invisible, comprising at once being and not-being, the origin of all things. All things are born of being. Being is born of not-being. All things originate from Tao. To Tao all things return. We have here a vague conception of the emanation philosophy. The creed of the faith is based on the virtue of passivity. Not to act, is the source of all power. The passive conquers. Passivity identifies one with Tao, and yields the strength of Tao to the believer. A certain flavor of Buddhism pervades this theory, and it may have had its origin in a previous knowledge of the Buddhistic creed by the philosopher; but it is very far below Buddhism in distinctness of statement and clearness of thought. Yet it is remarkable as the highest philosophical product of the Chinese mind.

If now we come to consider the ancient Aryan philosophies, it is to find ourselves in a new world of thought, a realm of the intellect that seems removed by a wide gulf from that occupied by the contemporary peoples of alien race. These philosophies are the work of two branches of the Aryans, the Hindu and the Greek, some brief account of whose systems of thought may be here given.

Of the peoples of the past only four can be said to have risen, in their highest thought, clearly above the level of mythology. These were the Chinese and the Hebrews, the Hindus and the Greeks; to whom may be added the pupils of the last, the Romans. But of these the first two named cannot be fairly said to have ever had a mythology. And of them the Hebrews originated no philosophy, while out of the countless millions of the Chinese race, with their constant literary cultivation, only one or two philosophers arose; and their systems of thought, perhaps devised under Buddhistic inspiration, have been allowed to decline into blank idolatry or unphilosophical scepticism. Far different was the case in India. There we find a connected and definite system of philosophy growing up, the outcome of the thought of a long series of Brahmanic priests, grounded in the childlike figments of mythology, but developing into a manly vigor of reasoning that has never been surpassed in the circle of metaphysical thought. It was a remarkable people with whom we are now concerned, — a people that dwelt only in the world of thought, and held the affairs of real life as naught. This world was to them but a temporary resting-place between two eternities, a region of probation for the purification of the soul. With the concerns of the eternities their minds were steadily occupied, and time was thrust aside from their thoughts as a base prison into which their souls had been plunged to purge them of their sins.

Their effort to solve the mystery of existence called forth an intricate and clearly thought-out conception of the organization of the universe, in which reason and imagination were intimately combined, — the latter, however, often so unchecked and extravagant as to reach heights of untold absurdity. The final outcome of this activity of thought was a philosophical system strikingly like that reached by the Egyptians, — a dogma of emanation and absorption, with intermediate stages of transmigration. But instead of the vapor-shrouded eternity of Egyptian thought, we here look into the past and the future of the universe through a lens of clear transparency.

We have now to deal with a thoroughly pantheistic doctrine of the universe, — the abundant fountain of all subsequent pantheism. In the beginning *Bráhmă* alone existed, — an all-pervading, self-existent essence, in which all things yet to be lay in the seed. This divine progenitor, the illimitable essence of deity, willed the universe into being from his own substance, created the waters by meditation, and placed in them a fertile seed, which developed into a golden egg. From this egg Bráhmă, the impersonal essence, was born into personal being as Brahmā, the creator of all things. We need not here concern ourselves with the many extravagances of the ardent Hindu imagination, that overlaid this conception and the subsequent work of creation with an endless array of fantastic adornments, but may keep to the central core of the Brahmanic philosophy. It will suffice to say that from the impersonal, thus embodied as the personal Brahmā, all things arose, — the heavens, the earth, and the nether realm, with all their countless inhabitants. All were emanations from the primal Deity, and all were destined to be eventually re-absorbed into this deity, so that existence should end, as it had begun, in Bráhmă alone. But with this descent from the infinite had come evil, or imperfection. Though a portion of the divine essence entered into all things, animate and inanimate, yet all things had become debased and impure. The one perfect being had unfolded into a limitless multitude of minor and imperfect beings. Such was the first phase of the mighty cycle of existence. The second phase was to be one of re-absorption, through which the multitude of separate beings would become lost in the one eternal being, and Bráhmă — who had never ceased to constitute the sole real existence — would regain his primal homogeneous state.

But divinity had become debased in the forms of men and animals, angels and demons. How was it to be purified, and rendered fit for absorption into the divine essence? In this purification lay the terrestrial part of the Hindu pantheism. To prepare for re-absorption into Bráhmă was the one duty of man. Attention to the minor duties of life detracted from this. Evil deeds still further debased the soul. The great mass of mankind died unpurified. But the divine essence in them could not perish. And in most cases it had become unfit to inhabit so high a form as the human body. Therefore it entered, after the death of men, into the bodies of various animals, into inanimate things, and even into the demonic creatures of the Hindu hell, in accordance with its degree of debasement. It must pass, for a longer or shorter period, through these lower forms ere it could be fitted to reside again in the human frame. And after having by purification passed beyond the human stage, it still had a series of transmigrations to fulfil, in the bodies of angels and deities, before it could attain the finality of absorption. To this ultimate, all Nature, from its highest to its lowest, was endlessly climbing. Everything was kindled by a spark of the divine essence, and all existence consisted of souls, in different stages of embodiment, striving upward from the lowest hell to the loftiest stage of divinity.

For these many manifestations of the one eternal soul there was but one road to purification. This lay through subjection of the senses, purity of life, and knowledge of the deity. Asceticism, mortification of the animal instincts, naturally arose as a resultant of this doctrine. The virtues of temperance, self-control, and self-restraint were the highest of human attainments. To reduce the flesh and exalt the soul was the constant effort of the ascetic, and to wean the mind from all care for the things of this life was the true path toward purification. Finally, knowledge of the deity could come only through a deep study of the Institutes of religion, rigid observance of its requirements, and endless meditation on the nature and the perfections of the ultimate essence, — the eternal deity. By thus giving the soul a steadily increasing supremacy over the matter that clogged and shadowed its pure impulses, in the end it would become utterly freed from material embodiment, and fitted to enter its final state of vanishment into the supreme. Just what this final state signified, whether the soul was or was not to lose all sense of individuality, is a question whose answer is not very clearly defined; and it is probable that the Hindu thinkers, bold as they were, shrank before this utterly insoluble problem, and left the final abyss uninvaded by their daring speculations.

It is a grand system of thought which we have here very imperfectly detailed, an extraordinary one to have been devised at so early a period, and by a people just emerging from barbarism into civilization. No higher testimony to the superiority of the Aryan intellect could be offered than to bring this clearly outlined cosmical philosophy into comparison with the confused, imperfect, and vapory conceptions of the Egyptian and the Chinese mind. It must be said, however, that it offers a conception of man's obligations as a citizen of the universe that has proved fatal to the national progress of the Hindu people. From the Brahman to the outcast, they have remained politically and socially dormant, their duties to the world to come dwarfing their duties to the world that is, and the realm of thought overlaying in their lives the realm of action. No heroes have risen to lead the Hindu people on the path to nationality or empire, for thinkers and workers alike have been lost in the shadow of a dream. The very thought of history-writing or history-making has not arisen among them ; and they have yielded with scarce a struggle to a long array of foreign conquerors, heedless of who ruled their bodies while their thoughts continued free.

The philosophy here described was, as we have said, the work of a long line of priestly thinkers, not of any great lawgiver of the race. In it we have the highest expression of the endlessly active Hindu intellect. At a later date, however, the names of several special thinkers emerge, each devising some variation in the details, yet none deviating from the basic principle of the system. The mystery of the origin of matter was left unaccounted for in the ancient Vedanta system ; and its actual existence was afterward denied, it being declared a mere illusion, arising from the imperfect knowledge of the soul. Kapila, the founder of the Sankhya school, attempted to overcome this difficulty by proclaiming the eternal existence of an unconscious material principle possessed of self-volition in regard to its own development. From it all matter had emanated, and into it all matter would be absorbed. By the side of this material principle existed a primal spiritual essence, manifold in its nature, and which from the beginning has entered into and animated matter. This spiritual unintelligence is endued with a subtile body consisting of intelligence (*buddhi*). The Sankhya deity is a compound of these three elements, — spirit, substance, and intelligence.

This scheme was followed by that of Patanjali, who considered the spiritual principle to be possessed of self-volition, and to exist separate from the co-eternal principle of matter. But the most striking of these speculative systems was that of Gautama, the founder of Buddhism, and the final great Hindu philosopher. This system was in the line of that of Kapila; but it carried the Hindu vein of thought to its utmost conceivable extension. It denied the existence of the soul as a substance. No spiritual essence pervaded the body. It held only certain intellectual attributes, which would perish with it. But the sum of each individual's good and evil actions (*Karma*) would survive, to migrate through other bodies, until the evil became eliminated, and only the good remained. As to the culminating stage of this process, the *Nirvana*, whether it signified the final extinction of evil and the vanishment of good, an utter and eternal nonentity, or embraced the conception of a conscious existence of the absolutely purified principle of good, — is a question that has been endlessly debated, and yet remains unsolved. The system made provision for the natural disappearance of evil; but the principle of good remained, and would not down at the command of thought. Probably the founder of the Buddhistic sect was as deeply lost as the Brahmanic philosophers in the abyss of infinity into which his daring conception had plunged. It is a depth by which all explorers have been baffled, and which the plummet of thought has ever failed to sound.

In regard to the manifold philosophies of Greece much

less need here be said. They are far better known to readers in general, and are to a large extent philosophies of the earth rather than schemes of the universe. The imagination of the Greeks was as bold and active as that of the Hindus; but it was far more under the control of the reasoning faculties, and is always subdued and artistic where that of the Hindus riots in the wildest extravagance. The Hindu philosophy directly emerged from the mythology of the Vedas and the sacrificial observances of the priests, and the steps of its evolution can yet be traced. The Greek philosophy had no relation to mythology. The gods of Greece had become so laden with earthly clay that they had ceased to be fit subjects for any but the vulgar belief when philosophy first showed its front on the Ionic shores. Thus the philosophy of Greece was a completely new growth. Cutting loose from all preceding thought, the Grecian intellect endeavored to construct a universe of its own, on the platform of what it saw and what it felt.

The various systems devised need be but rapidly run over, as they are more matters of ordinary knowledge than is the Hindu philosophy. The Ionic philosophers, Thales and his successors, endeavored to arrive at a conception of all existence from a study of the properties of physical substances, and the Pythagoreans from a like study of the properties of number. Next came the Eleatics, with their system of abstraction. Through the denial of the actuality of visible existence they arrived at a conception of *pure being*, — the basis of all appearance. Heraclitus followed, with his system of the *becoming*, — the incessant flow between finity and infinity, being and not-being. To these succeeded the Atomistic philosophers, to whom matter was the basis of being, and force the cause of movement. The philosophers here named were gradually advancing toward a theory of the universe; but it was a theory built up from the ground, rather than brought down from the infinite, as with the Hindus, — a scientific rather than an imaginative evolvement. As yet the idea of a deific principle had not appeared. This was devised by Anaxagoras, who placed a world-forming Intelligence by the side of matter. Yet the idea was only feebly grasped. This Intelligence existed but as a primary impulse, a moving force to set the universe in motion. The philosophic mind of Greece had not yet advanced to the grand outreach of Hindu thought.

This material phase of philosophizing was followed by the mental one of the Sophists and of Socrates. Cutting loose from the conception of matter as the basis of all things, they came to that of mind. The Sophists stood forth as the destroyers of the whole preceding edifice of thought, and Socrates as the originator of a new system of philosophy, in which the subjective replaced the objective, and mind subordinated matter. With him virtue and duty became the great principles of existence, thought was higher than matter, and morality superior to philosophy. He gave birth to no cosmology, but he turned the attention of man to a distinctively new field of speculation.

This was deeply worked by Plato, his great disciple, whose system of *Ideas* replaced the old systems of things, and with whom the supreme and all-embracing idea, the absolute *Good*, became the God, the divine creator and sustainer. Finally followed Aristotle, with his strongly scientific turn of mind and his highly indefinite metaphysical conception of the fluctuations between Potentiality and

Actuality, the variation from matter to form, from formless matter to pure or immaterial form. To these conceptions were added cosmological notions largely derived from the old mythology. But the value of the thought of Greece was not so much for its deductive as for its inductive labors. It tended constantly toward a scientific research into the basis of matter and mind, and never began by cutting loose from the actual, as in Hindu thought.

The mental acumen of these two highly intellectual branches of the ancient Aryans approached equality; but the real value of their work differed widely, mainly as a consequence of their different standpoints of thought. The speculations of the Greeks were based on observed facts, those of the Hindus on mythological fancies. As a consequence, the Greeks have worked far more truly for the intellectual advancement of mankind. If we come to glance at modern philosophy, a strikingly similar parallel appears. The Germans, the metaphysicians of the modern age, have inclined toward the Hindu line of pure deduction, and built vast schemes of philosophy with little more solid basis than the doctrine of emanation. The English and French, on the contrary, have developed the Greek line of science, and based their philosophies on observed facts. Their schemes do not tower so loftily as those of Germany, but they are built on the ground, and not on the clouds, and are likely to stand erect when the vast edifices of pure metaphysics have toppled over in splendid but irremediable ruin.

X.

THE ARYAN LITERATURE.

IT is not our intention to enter upon the task of a general review of the vast field of Aryan recorded thought, but merely to offer a comparative statement of the literary position of the several races of mankind, in evidence of the superiority of the Aryan intellect. Literary labor has been by no means confined to this race. Every people that has reached the stage of even an imperfect civilization has considered its thoughts worthy of preservation, its heroes worthy of honor, its deeds worthy of record. But so far as the intellectual value of literary work is concerned, the Aryans have gone almost infinitely beyond the remainder of mankind.

All early thought seems naturally to have flowed into the channel of poetry, with the exception of certain dry annals which cannot properly be classed as literature. This poetry, in its primary phase, appears to have been always lyrical. It was apparently at first the lyric of worship. This was followed by the lyric of action, and this, in its highest outcome, by the epic, — the combined and organized phase of the heroic poem. It is of interest to find that the Aryans alone can be said to have fairly reached the final stage of the archaic field of thought, the epic efforts of other races being weak and inconsequent, while almost every branch of the Aryan race rose to the epic literary level.

Of the antique era of the religious lyric little here need be said. We find it in the hymns of the Vedas and of the Zend-Avesta, in the early traditional literature of Greece, and in the ancient Babylonian hymns to the gods, some of which in form and manner strikingly resemble the Hebrew psalms. As to the second poetic period, that of the heroic song, or the record of the great deeds of the gods and demigods, little trace remains. Heroic compositions, as a rule, have ceased to exist as separate works, and have either become component parts of subsequent epics, or have vanished. As to valuable epic literature, however, it is nearly all confined within Aryan limits.

Modern research into the fragmentary remains of the ancient Babylonian literature has brought to light evidence of a greater activity of thought than we formerly had reason to imagine. And among the works thus recovered from the buried brick tablets of the Babylonian libraries are portions of a series of mythological poems of a later date than the hymns. These productions are considered to form part of an antique and remarkable poem, with a great solar deity as hero, — an epic centre of legend into which older lays have entered as episodes. It appears to have consisted of twelve books, of which we possess two intact, — the Deluge legend, and that of the descent of Istar into Hades; while part of a third exists, in which is described the war of the seven evil spirits against the moon. The Assyrians are supposed to have also had their epic, in imitation of this older work, and the Semiramis and Ninus of the Greeks are considered by M. Lenormant to have been heroes of this legendary circle of song. However that be, it cannot be claimed that either in poetic or artistic ability the Semitic mind displayed any exalted epic powers. So far as we are able to judge of this work from its scanty remains, it is devoid of all that we are accustomed to consider literary merit, and is full of hyperbolical extravagance.

Of the Semitic races, indeed, the Hebrews alone produced poetry of a high grade of merit. Of this Hebrew literature we shall speak more fully farther on, and it must suffice here to say that none of it reached the epic level. It is, as a rule, lyrical in tendency. Hebrew literature, however, is not without its heroic characters. We find them in Noah, Samson, David, Daniel, and others who might be named; but none of these were made heroes of song, but were dealt with in sober prose, — as we shall find later on was the fate of the heroes of Roman legend. The Hebrew intellect, indeed, was largely practical in its tendencies, its imagination was subdued, and though its literature contains many exciting legendary incidents, these are all couched in quiet prose, while its poetry fails to rise above the lyric of worship or of pastoral description. The nearest approach to an epic poem is the grand book of Job, of unknown authorship. The literature of Assyria, of which abundant relics are now coming to light, is yet more practical in character than that of the Hebrews, and resembles that of the Chinese in literalness. There is no poetry approaching in merit the elevated lyrical productions found in the Hebrew scriptures, and, like the Chinese, it is largely devoted to annals, topography, and other practical matters. The Semitic race as a whole appears to have been deficient in the higher imagination, though possessed of active powers of fancy. To the latter are due abundant stores of legend, often of a highly extravagant character; but we nowhere find an instance of those lofty philosophical conceptions, or of that high grade of epic song or dramatic composition, which are such frequent products of Aryan thought, and which indicate an extraordinary fertility of the imagination in the Aryan race.

Egypt produced little work of merit from a literary point of view. The religious literature consists of certain hymns of minor value, and the well-known "Ritual of the Dead." Similar to this is the "Ritual of the Lower Hemisphere." These ritualistic works can scarcely be called literary productions, and are marked by an inextricable confusion. So far as the display of intellectual ability is concerned, they are almost an utter void. In addition to its lyrics, Egypt has one work which has been dignified with the title of epic, though it should rather be viewed as an extended instance of those heroic legends whose confluence is needed to constitute a true epic production. It forms but the first stage in the production of the epic. This poem is credited to a scribe named Pentaur, and is devoted to a glorification of the deeds of Rameses II. in a war which that monarch conducted against the Cheta. He seems to have been cut off from his troops by the enemy, and to have safely made his way back to them. But the poem tells us that the mighty hero fell into an ambuscade of the Cheta, and found himself surrounded by two thousand five hundred hostile chariots. Invoking the gods of Egypt, the potent warrior pressed with his single arm upon the foe, plunged in heroic fury six times into their midst, covered the region with dead, and regained his army to boast of his glorious exploits. It is a bombastic and inartistic production; but such as it is it seems to have struck the Egyptian taste as a work of wonder, and has been engraved on the walls of several of the great temples of the land. The most complete copy of it is written on a papyrus now in the British Museum.

The remaining antique non-Aryan civilization, that of China, is utterly void of any epic productions, either in the ultimate or in the germ. The imagination necessary to work of this kind was wanting to the Chinese. Their decided practical tendency is abundantly shown in their close attention to annalistic history and to such subjects as geography, topography, etc. But no heroic legend exists, and but little trace of the devotional poetry with which literature begins elsewhere. The Confucian "Book of Odes," which contains all we possess of the antique poetry of China, is mainly devoted to the concerns of ordinary life. It has little of the warlike vein, but much of the spirit of peaceful repose. We are brought into the midst of real life, with domestic concerns, religious feeling, and family affection replacing the wild "outings" of the imagination which are shown in all the ancient Aryan literature. After the Confucian period Chinese song gained a somewhat stronger flight, and the domestic ballad was replaced by warlike strains and mythologic songs. But no near approach to epic composition was ever attained.

If now we enter upon Aryan ground we find ourselves at once upon loftier peaks of thought, and in a higher and purer atmosphere. Almost everywhere epic poetry makes its appearance at an early stage of literary cultivation as the true usher to the later and more practical branches of literature. These antique epic creations of the Aryans may be briefly summarized. As in philosophy, so in poetry, India and Greece take the lead; the Ramayana vying, though at a much lower level of art, with the Iliad of Greece. Of the two ancient epics of the Hindus, the *Ramayana* and the *Mahabharata*, the former is the older, while it is more the work of a single hand, and shows few signs of that epic confluence of legend which strongly characterizes the latter. And of the two, the Ramayana is the more mythological, the Mahabharata the more historical in character.

Legend credits northern India in these early days with two great dynasties of kings, known respectively as the Solar and the Lunar dynasties. The Ramayana describes the adventures of a hero of the solar race. Rama, the hero, is a lineal descendant of the god of the sun, and is himself adored as an incarnation of Vishnu. Everywhere in the poem we find ourselves on mythological ground, and the only historical indication it contains is that of the extension of the Aryan conquest southward toward Ceylon. The story describes the banishment of Rama from his hereditary realm and his long wanderings through the southern plains. His wife, Sita, is seized by Ravana, the giant ruler of Ceylon. Rama, assisted by Sugriva, the king of the monkeys, makes a miraculous conquest of this island, slays its demon ruler, and recovers his wife, the poem ending with his restoration to his ancestral throne.

The style of this poem is of a high grade of merit, and it takes a lofty rank among the works of the human imagination. In the first two sections there is little of extravagant fiction, though in the third the beauty of its descriptions is marred by wild exaggerations. It is evidently in the main the work of one hand, not a welding of several disjointed fragments. There are few episodes, while the whole latter portion is one unbroken narrative, and there is shown throughout an unvarying skill and poetical power and facility. It is credited to a single poet, Valmiki. This name signifies "white ant-hill," and it is very doubtful if it represents a historical personage. However that be, the Ramayana is a homogeneous and striking outcome of ancient thought.

The Mahabharata is a work of very different character. It is rather a storehouse of poetic legends than a single poem, and is evidently the work of many authors, treating subjects of the greatest diversity. It is of later date than the Ramayana, and more human in its interest, but is far below it in epic completeness and unity. Yet it is not without its central story, though this has almost been lost under the flood of episodes. It is the epic of the heroes of the lunar dynasty, the descendants of the gods of the moon, as the Ramayana is the heroic song of the solar race. Bharata, the first universal monarch, who brought all kingdoms "under one umbrella," has a lineal descendant, Kuru, who has two sons, of whom one leaves a hundred children, the other but five. The fathers dying, the kingdom is equitably divided among these sons, the five Pandavas and the hundred Kauravas. The latter grow envious, wish to gain possession of the whole, and propose to play a game of dice for the kingdom. The Pandavas lose in this strange fling for a kingdom; but the Kauravas agree to restore their cousins to their share in the throne if they will pass twelve years in a forest and the thirteenth year in undiscoverable disguises. This penance is performed; but the Kauravas evade their promise, and a great war ensues, in which the Pandavas ultimately triumph. Whether this war indicates some actual event or not, is questionable; but this part of the work is well performed, the characters of the five Pandavas are finely drawn, and many of the battle-scenes strikingly animated.

But this main theme forms but a minor portion of the work. It is full of episodes of the most varied character, and contains old poetical versions of nearly all the ancient Hindu legends, with treatises on customs, laws, and religion, — in fact, nearly all that was known to the Hindus outside the Vedas. The main story is so constantly interrupted that it winds through the episodes "like a pathway through an Indian forest." Some of these episodes are said to be of "rare and touching beauty," while the work as a whole has every variety of style, dry philosophy beside ardent love-scenes, and details of laws and customs followed by scenes of battle and bloodshed. Many of the stories are repeated in other words, and the whole mass, containing more than one hundred thousand verses, seems like a compilation of many generations of Hindu literary work. Yet withal it is a production of high merit and lofty intellectual conception.

In regard to the Persian branch of the Indo-Aryans, it yields us no ancient literary work in this exalted vein. That considerable legendary poetry existed we have good reason to believe; but it does not seem to have centred around a single hero, as elsewhere, but to detail the deeds of a long series of legendary kings, many of whom were undoubtedly historical personages. It was late in the history of the Persians when these legends became condensed into a single work, the celebrated *Shah Namah* of

Firdusi, which forms, as Malcolm observes, "deservedly the pride and delight of the East." It professes to be but a versified history of the ancient Persian kings, from the fabulous Kaiomurs to the fall of the second empire under Yezdijird. But no trace remains of the documents employed by the poet, while his work is to so great an extent legendary that it has all the elements of the epic except that of a central hero. The work itself displays the highest literary skill and poetical genius, and, as Sir John Malcolm remarks, "in it the most fastidious reader will meet with numerous passages of exquisite beauty." The narrative is usually very perspicuous, and some of the finest scenes are described with simplicity and elegance of diction, though the battle-scenes, in which the Persians most delight, are by no means free from the Oriental besetting sin of hyperbole.

Of the epic poetry of Greece, and particularly the great works attributed to Homer, little here need be said. The *Iliad* and *Odyssey* are too well known to readers to need any description. Modern research has rendered it very probable that these works, and the Iliad in particular, are primitive epics in the true sense, being condensations of a cycle of ancient heroic poetry. The antique Greek singers were not without an abundant store of stirring legends as subject-matter for their songs. These legends have become partly embodied in poetry, partly in so-called history; and in them mythology, history, and tradition are so mingled that it is impossible to separate these constituents and distinguish between fact and fancy. But of all the legendary lore of the Greeks, that relating to the real or fabulous siege of Troy seems most to have roused the imagination of the early bards, and brought into being a series of the most stirring martial songs. These as a rule centred around the deeds of one great hero, Achilles, the scion of the gods, the invulnerable champion of the antique world.

Little doubt is entertained by critics that the Iliad contains the substance of a number of ancient lays devoted to this one attractive subject. But if so, there can scarcely be a doubt that these lays were fitted by a single skilful hand into the epic framework of the Homeric song. We may as well seek to divide Shakspeare into a series of successive dramatists as to break up Homer into a cycle of antique poets. Men of his calibre do not arise in masses, even in the land of the Hellenes; and though there can be little question that older material made its way into the Iliad, there can be as little question that it was wrought into its present form by one great genius, and fitted by one skilful hand into the place which it occupies. Another theory offered is that the nucleus of the poem and a portion of its incidents are the work of a single great poet, while episodes of other authorship were worked into it at a later period. But a more probable supposition would seem to be that Homer, like Shakspeare, dealt with heroic legends of earlier origin, ancient ballads whose substance was worked into the nucleus of the poem by that one great genius whose vital intellect inspirits the whole song. This would explain at once the discrepancies that exist between the subject and handling of the several cantos, and the considerable degree of unity and homogeneity which the poem as a whole possesses. It need scarcely here be said that the Iliad stands at the head of all epic song, alike in the manner of its evolution, the lofty poetic genius which it displays, and the exquisite beauty of its versification. As compared with the Hindu epics, it displays the artistic moderation of Greek thought in contrast with the unpruned exuberance of the Oriental imagination. Even the gods which crowd its pages are as human in their lineaments as a Greek statue, and we are everywhere introduced to the society of actual man, with his real passions, feelings, and sentiments, instead of to a congeries of phantasms whose like never drew breath in heaven, earth, or sea.

The Odyssey has been subjected to criticism of the same character, and with like indefinite results. There can be no doubt that here also we have to do with one of the favorite heroes of Greek legend, — the wise, shrewd, hard-headed old politician Ulysses, in contrast with the fiery Achilles, uncontrollable alike in his fury and his grief. They are strongly differentiated types of character, both to be found in the mental organization of the Greek, and perhaps chosen from an involuntary sense of their fitness. We need not here follow Ulysses in his wanderings and his strange adventures by land and sea. They simply indicate the conception of the ancient Greek mind, yet firmly held in mythologic fetters, of the conditions of the world beyond its ken. Yet a considerable change had taken place in the ruling ideas between the dates of the two poems. The turbulent Olympian court of the Iliad has almost disappeared in the Odyssey, and Zeus has developed from the hot-tempered monarch of the Iliad into the position of a supreme moral ruler of the universe. If both poems are the work of one hand, which is now strongly questioned, the poet must have passed from the ardent and active youth of the Iliad to the reflective era of old age and into a period of developed religious ideas ere he finished his noble life-work with the Odyssey.

Of the remaining epic work of Greece nothing need be said. The true epic spirit seems to have died with Homer: and though many heroic poems were afterward produced, they lack the lofty poetic power of the ancient Muse. But one work need be named here, the *Theogony* of Hesiod, as at once partly an epic poem, and partly a mythological record. To a certain extent it may be classed with the Icelandic Eddas and the Persian cosmogony; though the scheme which it presents is less connected and complete, and it cannot lay the same claim to the title of a philosophy of mythology. On the other hand, it details many stirring scenes, and its description of the battles between Zeus and the Titans has an epic power which approaches that of Milton's story of the war on Heaven's plains.

The epic poetry of Rome may be dismissed with a few words. That the Romans possessed the vigor of imagination and the boldness and sustained energy of conception necessary to work of this description, is sufficiently attested by the *Æneid* of Virgil. But it is with a native epic growth that we are here concerned, not with a secondary outcome of Greek inspiration. A study of ancient Roman history reveals the fact that abundance of epic material existed. This history is in great part a series of legends, many of which are doubtless prose versions of old heroic lays. Cicero remarks that "Cato, in his *Origines*, tells us that it was an old custom at banquets for those who sat at table to sing to the flute the praiseworthy deeds of famous men."[1] He further regrets that these

[1] Quaestiones Tuscul. iv. 2.

lays had perished in his time. Other writers give similar testimony; and it is highly probable that the stories of the warlike deeds of Horatius, Mucius, Camillus, etc., were largely poetic fictions, designed to be sung in the halls of the great nobles of these clans. We find here no clustering of legend round the names of single heroes, as in ancient Greece. The scope of Roman thought lay below the level of the demigods. It was practical throughout, and permitted but minor deviations from the actual events of history. Thus Roman legend is more in the vein of that of Persia, which was spread over a long line of fabulous kings, instead of concentrating itself around a few all-glorious champions. Rome, however, produced no Firdusi to embalm its legends in the life-like form of song. Yet the history of Livy may almost be called an epic in prose. It is the nearest approach which Rome made to a national epic, and prose as it is, the great work of Livy deserves to be classed among the heroic epics of the world.

It is in strong confirmation of the intellectual energy of the Aryans to find that the remaining and more barbaric branches of the race, equally with the Greeks and Hindus, produced their epics of native growth. And it is of interest to find that the Teutonic and Celtic epic cycles display the true epic condition of the concentration of a series of heroic lays around one great national hero. With the Teutonic people a native Homer arose to give epic shape to the floating lays of the past. This cannot be affirmed of the Celts, whose ancient heroes owed their final glory to foreign hands.

The Germans possess more than one collection of antique lays, such as the poem of *Gudrun*, and the *Helden-*

buch, or Book of Heroes. But it is to the *Nibelungen-lied* that they proudly point as a great national epic, the outgrowth of their heroic age. Nor is this pride misplaced. The song of the Nibelung is undoubtedly a great and noble work, unsurpassed in the circle of primitive warlike epics except by the unrivalled Iliad. It is full of the spirit of the old German lays, such as Tacitus tells us the Germans of his time composed in honor of their great warriors. It is full also of mythological elements, to such an extent that it is difficult to discriminate between the deific and the human origin of its heroes. In its central hero, Siegfried, the Achilles of the song, and in the heroic maiden Brunhild, we undoubtedly have mythological characters. But in others, such as Etzel and Dietrich, can be traced such well-known historical personages as Attila, the leader of the Huns, and Theodoric, the Gothic king. Siegfried and Brunhild appear in other legends besides those of the Nibelung, and we find the former in the Volsung lay of the Eddas as Sigurd, who fought with the dragon Fafnir for the golden hoard. This golden hoard is a moving impulse in the Teutonic legendary cycle. Siegfried has become the possessor of the enchanted treasure of the Nibelungs, and, like Achilles, has been made invulnerable, except in a spot between his shoulders, which replaces the heel of Achilles.

But the hoard of gold is a secondary motive in the Nibelungen-lied. Its mythologic fiction has almost vanished, and has been replaced by human motives, human passions, and human deeds. Man has dwarfed the gods in this outcome of German thought. It is the truly human passion of jealousy, the hot rivalry of the two queens, Brunhild and Kriemhild, and the bitter thirst of the latter for revenge, that carry us through its stirring epic cycle of treachery, war, and murder. There is nothing in the whole circle of song more terrible than the *finale* of this vigorous poem, the pitiless battle for vengeance in the blood-stained banquet-hall of the Huns. Of the name of the poet who shaped the old ballads into the enduring form of the Nibelungen-lied we have no more than a conjectural knowledge. This work was apparently done about the year 1200; but the lays themselves perhaps reach back to the fifth or sixth centuries. The epic work was done by a master-hand, who has moulded the separate songs, sagas, and legends into a well-harmonized single poem with a judgment and ability that shows the possession of a vigorous genius.

The Nibelungen-lied is not a courtly poem. It is full of the rudeness and passion of a barbaric age, though the conditions of Middle-Age society, with its combined cruelty and chivalry, and the sentiment of the age of the Minnesingers, have not been without their effect in softening the spirit of the older lays, and in giving a degree of poetic splendor to the crude boldness of archaic song. It falls far below the Iliad in all that constitutes a great work of art, yet it is instinct with a fervent imagination, a fiery energy, and a truly epic breadth of incident. Its descriptive power, the fine characterization of its personages, and the skilful handling of the plot, indicate both an age of considerable literary culture and a high degree of poetic genius in the narrator, while the Teutonic spirit is shown in its deep feeling for the profound and mysterious in human destiny. Opening with a calm and quiet detail of peaceful incidents, we soon find the poem plunging into the abyss of jealousy, rivalry, murder, and all the fiercer passions. The hand of the assassin finds the vulnerable spot in Siegfried's body, the fatal spot left unbathed by the magic dragon's blood, and he falls a victim to Brunhild's relentless hate. From this point onward the poem gathers force as it flows, until it sweeps with the fury of a mountain-torrent toward its disastrous *finale* in the terrible retribution exacted by the hero's vengeance-brooding wife. The death-dealing spirit of ancient tragedy finds its culmination in the story of awful bloodshed in which the murderous Hagen and his companions meet their deserts at the court of the Huns. The terrible energy with which the poem closes finds nothing to surpass it in the most vigorous scenes of Homer's world-famous works.

One more poem of epic character, the product of the Teutonic Muse, may be here mentioned, — the most archaic and barbarous of all epic songs. This is the primeval English epic, the poem of *Beowulf*, — the work of the Anglo-Saxons in their days of utter barbarism and heathenism, probably before they left their home on the Continent to fall in piratical fury on England's defenceless shores. We have here no chivalry, no sentiment, no softness. All is fierce, rude, and savage. The superstitions of an age of mental gloom form the web of the poem, which is shot through and through with the threads of mythologic lore. It is, as Longfellow remarks, "like a piece of ancient armor, — rusty and battered, and yet strong." The style is of the simplest. The bold metaphorical vein of later Anglo-Saxon poetry is wanting; the poet seems intent only on telling his story, and has no time for episodes and metaphors. Yet Beowulf is the far-off progenitor of the knight-errant of chivalry; and the song is such as the uncultured, yet vigorous-minded, bards of the heathen Saxons might have sung in the rude halls of half-savage thanes,—ale-quaffing, stool-seated Berserkers, listening in the light of flaring and smoking torches to the stirring lay of human prowess and magic charms.

We are told how Beowulf, the sea Goth, fought unarmed with Grendel the giant, and destroyed the monster, after the latter had slain scores of beer-drunken doughty Danes in the great hall of King Hrothgar the Scylding. There succeeded a terrible fight in the dens whither Beowulf had followed the Grendel's mother, a witch-like monster. Here he slew dragons and monsters that blocked his way; and after a hard struggle with the grim old-wife, seized a magic sword which lay among the treasures of her dwelling, and "with one fell blow let her heathen soul out of its bone house."[1] To this strongly told bit of heathen lore are added eleven more cantos, relating the deeds of the sea-king in his old age, when he fought with a monstrous fire-drake which was devastating the land. He killed this creature, and enriched the land with the treasure found in its cave; yet himself died of his wounds.

Here again we have the magic treasure of Teutonic lore, destined to be fatal to its possessor, as the Nibelung hoard was to the hero Siegfried. It is undoubtedly an outgrowth of Northern mythology, and perhaps had its origin in the treasures of the dawn or of the summer of ancient Aryan myth. As an epic, the poem possesses much merit. It is highly graphic in its descriptions, while the story of its battles, its treasure-houses, the revels and songs in the kings' halls, and the magical incidents with which the poem is filled, are told with a minuteness that brings clearly before our eyes the life of a far ruder age

[1] Longfellow, Poets and Poetry of Europe, p. 4.

than is revealed by any other extended poem. As Longfellow says, "we can almost smell the brine, and hear the sea-breezes blow, and see the mainland stretch out its 'sea-noses' into the blue waters of the solemn main." This rude old song, so fortunately preserved, yields us striking evidence of the intellectual vigor of the fathers of the English race.

The Celtic Aryans have been quite as prolific as any other branch of the race; and though they present us with no completed epic, they have preserved an abundance of those heroic tales which form the basis of epic song. While the Germans of the Continent and the Saxons of England were plunged in the depths of barbarism, the Irish Celts manifested a considerable degree of literary activity, and produced works on a great variety of subjects, whose origin can be traced back to the early centuries of the Christian era. Among these were numerous heroic legends which centred around two great traditional champions of the past. One of these cycles of epic lays, whose heroes have almost vanished from the popular mind, relates the deeds of a doughty hero, Cúchulaind, of whose mighty prowess many stirring stories are told. The central tale is the *Tain Bo Cuailnge*, or the "Cattle Spoil of Cualnge," which tells how Cúchulaind defended Ulster and the mystic brown bull of Cualnge single-handed against all the forces of Queen Medb of Connaught, the original of the fairy-queen Mab. Around this vigorously told story cluster some thirty others, descriptive of the deeds of the hero Cúchulaind, of Medb the heroine, and of many great champions of the past. As a whole, it forms a complete epic cycle, and needed only the shaping and pruning hand of some able poet to add another to the national epics of the world. These legends, as they exist now, are in twelfth-century manuscripts, of mixed prose and verse; but for their origin we must go back to the vanished bards of many centuries preceding.

In addition to this epic cycle of heroic song, the Irish have the fortune to possess another, equally extensive, and of much more modern date, — the story of Finn, the son of Cumall, who is still a popular hero in Ireland, though his predecessor has long been forgotten. Finn and the Fennians may have had a historical basis, though there can be very little of the historical in the stories relating to them, with their abundance of magical incidents and extraordinary adventures. The Fennian tales probably only began to be popular about the twelfth century, and new ones continued to appear till a much later period, one of them being as late as the eighteenth century. These legends are very numerous, and they may claim to have found their epic poet in a bard of alien blood; for it seems certain that the heroes of both these cycles of songs were popular in the Highlands of Scotland, and that Macpherson's *Ossian*, though doubtless due, as a poem, to his own mind, contains elements derived by him from the popular Highland heroic lore. Ossian is Oisin, the son of Finn, while the hero himself is represented in Fingal; and characters of both the Irish legendary cycles are introduced. Much as the statement of Macpherson concerning the origin of this poem has been questioned, it may have equal claim to the title of a naturally evolved epic as the Nibelungen-lied or the Iliad. For in none of these cases are we aware to what an extent the final poet manipulated his materials, or how greatly he transformed the more ancient lays and legends.

The Welsh division of the Celts seems to have been nearly as active as the Irish in literary work, and produced its distinct epic cycle in the heroic lays of King Arthur, — the popular hero of the age of chivalry and of modern English epic song. This hero of fable, with his Round Table of noble knights, and the deeds of the enchanter Merlin, was first introduced to Middle-Age Europe in the fabulous British history of Geoffrey of Monmouth, written early in the twelfth century. The Arthurian legends yielded nothing that we can call an epic, but they gave inspiration to a marvellous series of rhymed romances, the work of the French Trouvères. The French, however, were not without a native hero of romance of older date in their literature than the Arthur myths. This was their great King Charlemagne, who, with his twelve peers, formed the theme of an interminable series of *Chansons de Gestes*, or legendary ballads, in which the epic spirit became diffused through a wide range of rude and magical romance. King Arthur succeeded Charlemagne as a popular hero at a period of more culture and softer manners, and the poems of which he and his knights form the heroes are the finest in that tedious series of magical romances with which the Trouvères and their successors deluged the literature of the chivalric age, until they finally sank into utter inanity, and were laughed out of existence by Cervantes in his inimitable satire of Don Quixote.

In this review of the early poetry of the Aryans there is one branch of the race yet to be considered, and one remaining epic to be described. The Slavonians have not been without their literary productions, though none of their poetry has reached the epic stage. But the contiguous Finns, whom we have viewed as nearly related in race to the Slavonic Aryans, have evolved an epic poem of some considerable merit, and of interest as the latest work of this character to come into existence in the primitive method. Its elements long existed among the Finnish people as a series of heroic legendary ballads, the work of arranging which into a connected epic form was due to Dr. Lönnrot, of Helsingfors, who collected from the lips of the peasantry, and published in 1835, the epic production now known as the *Kalevala*, the "Home of Heroes." These legends belong mainly to the pre-Christian period of Finnish culture. They centre, in true epic style, round the hero Wainamoinen, whose deeds, with those of his two brother heroes, form the theme of a series of connected lays, which fall together into a poem almost as homogeneous as the Iliad. It is a work instinct with mythology. It opens with a myth of the creation of the universe from an egg, and is full of folk-lore throughout. The heroes of Kaleva, the land of happiness, bring down gifts from Heaven to mortals, and work many magic wonders. Yet they mingle in the daily life of the people, share their toils, and enter into their rest. They are, as Mr. Lang says, "exaggerated shadows of the people, pursuing on a heroic scale, not war, but the common business of peaceful and primitive men." Yet the poem is not without its warlike element, — in the struggle of the heroes of Kaleva with the champions of Pohjola, the region of the frozen North, and of Luonela, the land of death. It ends, after many vicissitudes, in the triumph of Wainamoinen and his followers over their foes. Of the merits of this poem, Max Müller remarks: "From the mouths of the aged an epic poem has been collected, equalling the Iliad in length and completeness, — nay, if we can forget for the moment all that *we* in our youth learned to call beautiful, not less beautiful." In metre and style it resembles Longfellow's "Hiawatha," which imitates it with some exactness.

Though the Slavonic people have produced no heroic epos of this completeness, they are not without their heroic poetry. The success attained by Dr. Lönnrot in studying the popular poetry of Finland has led to like efforts in Russia, with very marked results. Two great collections of the epic lays of the Russian people now exist, — that published by P. N. Ruibnikof in 1867; and that of P. R. Kiryeevsky, which is not yet completed. These lays were collected from the lips of the Russian peasantry, the whole country being traversed by the ardent explorers in their indefatigable search for the old songs of the Slavonic race. The *Builinas*, or historic poems, thus rescued from oblivion seem naturally to fall into several cycles, each with its distinct characteristics. Of these the most archaic lays deal with the "Elder Heroes," and are evidently of mythologic origin. Closely connected with these in character is the cycle named after Vladimir the Great. This is the epos of the "Younger Heroes," — the ancient paladins of the country, like those of the Charlemagne and Arthur legends. The third is known as the Novgorod cycle, and deals with the remote era of historic Russia. The fourth is the Royal or Moscow cycle, and has the personages of actual history for its heroes.

These Russian songs show no tendency to centre round any single hero, and thus offer no opportunity for their concentration into a single connected poem. In the history of national epic poetry, in fact, we seem to distinguish two distinct lines of development. One of these is that pursued by Persia, Rome, and Russia, in which no single hero has concentrated the attention of singers, and the flow of song takes in a long succession of fabulous and historical champions. The other is that pursued by the remaining Aryans, in which song centred itself around one or a few great warriors, mostly of mythological origin, and the series of songs naturally combined into a connected narrative. This is the more archaic stage of the two, or perhaps the one that indicates the most active imagination, and it is the one to which all the naturally evolved epic poems of the world are due.

The production of heroic poetry by the Aryan peoples by no means ceased with their stage of half-barbaric development. Numerous valuable epic poems have been produced in the age of civilization; but of these we need say nothing, as they are secondary products of the human mind, and not the necessary outcome of mental evolution. They are only of value to us here as evidences of the continued vigor of the Aryan imagination. One only of these presents any of the characteristics of a naturally evolved work. This is the great poem of Dante, the *Divina Commedia*, in which the Middle-Age mythology of the Christian Church has become embodied in song, the record of a stage of thought which can never be reproduced upon the civilized earth. The Inferno of Dante is the mediæval expression of a succession of extraordinary conceptions of the future destiny of the soul. These are of strict Aryan origin, since all non-Aryan nations have had very vague conceptions of the punishment of the wicked. The extreme unfoldment of the hell-idea we owe to the Hindu imagination, and a less exaggerated one to that of Persia. It would be difficult to conceive of a more grotesquely extravagant series of future tortures than those of the Buddhistic hell. These ideas have been carried by the Buddhists to China, while they gave the cue to Mohammed and instigated the hell of the Koran. Their final product is the hell of mediæval Europe, and they have attained poetical expression in Dante's Inferno. We may therefore fairly class this poem with the primitive epics of mankind, as it gives poetic expression to a stage of human culture and a natively evolved series of mythical conceptions which have died out with the advance of civilization, but which were as essential elements of thought-development as the worship of mythical deities and the admiration of heroic demigods.

We have given considerable attention to the development of Aryan epic poetry from the evidence which it presents of the distinctly superior character of the Aryan imagination to that of the other races of mankind. None of these can be fairly said to have reached the epic level of thought. The Aryans have continuously progressed beyond this level. But the steps of this progression can here but concisely be indicated. The epic spirit in ancient Greece unfolded in two directions, one producing the imaginative historical narrative, the other giving rise to the drama. The former of these in that actively intellectual land quickly developed into history in its highest sense, yielding the rigidly critical and philosophical historical work of Thucydides. The latter as quickly gave rise to a succession of the noblest dramatic productions of mankind, those of the three great tragedians of Greece. Elsewhere in the ancient world the course of development was much the same. Rome produced no native drama of literary value, but in historic production it rivalled the best work of Greece, passing from the half-fabulous historical legends of Livy to the critical production of Tacitus. In this respect practical Rome was in strong contrast to imaginative India, in which land history remained undeveloped, while a drama of considerable merit came into existence.

If now we consider the unfoldment of modern European literature, it is to find it pursue a somewhat different channel, and reach results not attained in ancient times. The rhymed romance of chivalry was the direct outgrowth of the epic spirit in mediæval Europe, and was accompanied by metrical histories as fabulous as the romance. In their continued development these two forms of literature deviated. The history of fable gradually unfolded into the history of fact. Prose succeeded verse, and criticism replaced credulity. The rhymed romance, on its part, developed into the prose romance, and lost more and more of its magical element, until it fully entered the region of the possible. It still continued tedious and extravagant, but had got rid of its old cloak of mythology.

Ancient fiction reached a stage somewhat similar to this, though not by the same steps of progress. In the later eras of Greece romantic fictions appeared, comprising pastoral, religious, and adventurous tales similar to those which were the ruling fashion of a few centuries ago in Europe. But there was little trace of the allegory, which became such a favorite form of literature with our forefathers. In India this development stopped at a lower stage, that of fable and fairy lore. But in this field the active Hindu imagination produced abundantly, and directly instigated the Persian and Arabian magical literature. Through the latter its influence entered modern Europe. Collections of the Hindu tales were extant in the Middle Ages, and from them seems to have directly outgrown the short novel or tale, which attained such popularity and reached its highest level of art in the Decameron of Boccaccio.

But in more modern times the imaginary narrative has passed onward to a far higher stage than it attained in the ancient period, and has yielded the character-novel of our own day, — a literary form in which the combined imagination and reason of the Aryan mind have gained their loftiest development. The novel is the epic of the scientific and reflective era. It has cast off the barbaric splendor of the mantle of verse and of magical and supernatural embellishments, and has descended to quiet prose and actual life conditions. It has left the heroic for the domestic stage. It has replaced the outlined characters of the epic by critical dissections that reveal the inmost fibres of human character. The stirring action of the epic has in it been replaced in great part by reflection and mental evolution. It forms, in short, the storehouse into which flows all the varied thought of modern times, there to be wrought into an exact reproduction of the physical, social, and mental life of man.

The modern drama unfolded at an earlier date than the novel. But its evolution was a native one only in Spain and England. Elsewhere it was but an imitation of the drama of the ancient world. It attained its highest level in the works of Shakspeare, which indeed prefigured the modern novel in the critical exactness and mental depth of their character-pictures and in the reflective vein which underlies all their action. As complete reproductions of intellectual man, and dissections of the human understanding in its every anatomical detail, they probably stand at the highest level yet reached by the powers of human thought. The remaining outgrowth of epic narrative, that of prose history, has likewise attained a remarkable development in modern times, and has become as philosophical and critical as the narrative of ancient times, with few exceptions, was crude, credulous, and unphilosophical.

If an attempt be made to compare the literary work of the non-Aryan nations in these particulars with the Aryan productions, it will reveal a very marked contrast between the value of the two schools of thought. Nothing need be said of the fictitious or historical literature of the ancient non-Aryan civilizations. It lay in intellectual power very far below the level attained by Greece. The only important literary nation of modern times outside the Aryan world is China. In the making of books the Chinese have been exceedingly active, and their literature is enormous in quantity; the Europeans scarcely surpass them in this respect. But in regard to quality they stand immeasurably below the Aryan level.

Though China has produced no epic poem, it has been very prolific in historical and descriptive literature and in what is called the drama and the novel. Yet in its historical work it has not gone a step beyond the annalistic stage. The idea of historical philosophy is yet to be born in this ancient land. As for tracing events to their causes, and taking that broad view of history which converts the consecutive detail of human deeds into a science, and displays to us the seemingly inconsequential movements of nations as really controlled by necessity and directed by the unseen hand of evolution, such a conception has not yet entered the unimaginative Chinese mind.

As regards the Chinese drama and novel, they are utterly unworthy of the name. Character-delineation is the distinctive feature of the modern novel, and of this the novel of China is void. It consists mainly of interminable dialogues, in which moral reflections and trifling discussions mingle, while the narrative is made tedious by its many inconsequential details. The stories abound in sports, feasts, lawsuits, promenades, and school examinations, and usually wind up with marriage. There is abundance of plot, but no character. Their heroes are paragons of all imaginable virtues, — polished, fascinating, learned; everything but human. The same may be said of the Chinese drama. It is all action. Reflection and character-analysis fail to enter. There are abundance of descriptions of fights and grand spectacles, myths, puns, and grotesque allusions, intermingled with songs and ballets. The plot is sometimes very intricate, and managed with some skill; but often the play is almost destitute of plot, though full of horrible details of murders and executions. Fireworks, disguised men, and men personating animals, are admired features of those strange spectacles; but as for any display of a high order of intellectuality, no trace of it can be discovered in the dramatic or fictitious literature of this very ancient literary people.

There is no occasion, in this review, to consider all the many divisions into which modern Aryan literature has unfolded. There is, however, yet another of the ancient and naturally evolved branches of literature to be taken into account. We have said that the general course of poetic development seems to have been from the religious through the heroic lyric to the epic. But lyric poetry continued its development, accompanying and succeeding the epic. It has indeed come down to our own times in a broad flood of undiminished song. It is with the lyric, truly so called, that we are here concerned, — the poetry of reflection, the metrical analysis of human emotion and thought, in contrast with the poetry of action. To this may be added the poetry of description, of the love-song, and of the details of common life, with all their numerous varieties.

In this field of literature alone the other races come more directly into comparison with the Aryan. Prolific as every branch of the Aryan race has been in lyric song, the remaining peoples of civilized mankind have been little less so, and in this direction have attained their highest out-reach of poetic thought. The Hebrews specially excelled in the lyric. In the poem of moral reflection and devotion, in the delineation of the scenes and incidents of rural life, and in the use of apposite metaphor, they stand unexcelled, while in scope of sublime imagery the poem of Job has never been equalled. This poetry, however, belongs to a primitive stage of mental development, — that in which worship was the ruling mental interest of mankind. The intellect of man had not expanded into its modern breadth, and was confined to a narrow range of subjects of contemplation.

At a later period the Semitic race broke into a second outburst of lyric fervor, — that of the Arabians in their imperial era. But this failed to reach any high standard of intellectual conception. Their poems were largely devoted to love and eulogy; and while they had the same metrical harmony as their direct successors, the works of the Troubadours and the Minnesingers, they, like these, were largely void of thought, and lacked sufficient vitality to give them continued life. In China, again, we find a very considerable development of non-Aryan lyric song, coming down from a very early period of the nation. And these lyrics have often much merit as quiet pictures of life; but it cannot be claimed that they show any lofty intellectual power. For the highest development of the lyric, as of every form of literary work, we must come to the Aryan world, where alone thought has climbed and broadened, reaching its highest level and its widest outlook, and sinking to its profoundest depth of analysis of the mental universe. So far as literature embodies the powers of the human intellect, it points to the Aryan development as supremely in advance of that of the other races of mankind.

XI.

OTHER ARYAN CHARACTERISTICS.

IT is necessary, in continuation of our subject, to consider the comparative record of the Aryan and the other races of mankind in respect to the development of art, science, mechanical skill, and the other main essentials of civilization. In doing so, certain marked distinctions make themselves apparent, and it seems possible to draw broad lines of demarcation between the principal races. If we consider the Negro race from this point of view, it is to find a lack of energy both physical and mental. Nowhere in the region inhabited by this race do we perceive indications of high powers either of work or thought. No monuments of architecture appear; no philosophies or literatures have arisen. And in their present condition they stand mentally at a very low level, while physically they confine themselves to the labor absolutely necessary to existence. They neither work nor think above the lowest level of life-needs; and even in America, under all the instigation of Aryan activity, the Negro race displays scarcely any voluntary energy either of thought or work. It goes only as far as the sharp whip of necessity drives, and looks upon indolence and sunshine as the terrestrial Paradise.

The record of the Mongolian race is strikingly different. Here, too, we find no great scope or breadth of thought, but there is shown a decided tendency to muscular exertion.

For pure activity of work the Mongolians have been unsurpassed, and no difficulty seems to have deterred them in the performance of the most stupendous labors. The Aryans have never displayed an equal disposition to hand-labor, — not, however, from lack of energy, but simply that Aryan energy is largely drafted off to the region of the brain, while Mongolian energy is mainly centred in the muscles. The Aryan makes every effort to save his hands. Labor-saving machinery is his great desideratum. The Mongolian, with equal native energy, centres this energy within his muscles, while his brain lies fallow. The Chinese, for instance, are the hardest hand-workers in the world. The amount of purely physical exertion which they perform is nowhere surpassed. The productiveness of their country, through the activity of hand-labor alone, is considerably superior to that of any other country not possessed of effective machinery. But in regard to thought they exist in an unprogressive state. Little has been done by the brain to relieve the hand from its arduous labor. Chinese thought is mainly a turning over of old straw. The land is almost empty of original mental productions.

If we consider the record of the Mongolians of the past the same result appears. They have left us monuments of strenuous work, but none of highly developed thought. China, the most enlightened of Mongolian nations, has an immense ancient literature, but none that can be compared with Aryan literature in respect to display of mental ability. Its highest expression is its philosophy, and that, in intellectual grasp, is enormously below the contemporary philosophy of India. But in respect to evidences of muscular exertion it has no superior. The Great Wall of China far surpasses in the work there embodied any other single product of human labor. Yet it is in no sense an outcome of advanced thought. It is the product of a purely practical mind, and one of a low order of intelligence, as evidenced by the utter uselessness of this vast monument of exertion for its intended purpose. The Great Canal of China is another product of a purely practical intellect. Every labor performed by China has a very evident purpose. It is all industrial or protective. There are no monuments to the imagination. Yet the lack of mental out-reach has prevented any great extension of labor-saving expedients. At long intervals, during the extended life of the nation, some useful invention has appeared, — such as that of the art of printing. Yet for much more than a thousand years this art has remained in nearly its original stage, while in Europe, during a considerably shorter period, it has made an almost miraculous advance. Among the few illustrations of non-practical labor in China are its pagodas, which seem like the playthings of a rudimentary imagination when compared with the architectural monuments of Europe.

If now we review the products of the American aborigines, whose closest affinities are certainly with the Mongolians, we arrive at a similar conclusion. There is evidence of an immense ability for labor, but of no superior powers of thought. The quantity of sheer muscular exertion expended on the huge architectural structures and the great roads of Peru, the immense pyramids of Mexico, and the great buildings of Yucatan, is extraordinary. The huge mounds erected by the ancient dwellers in the Mississippi valley are equally extraordinary, when we consider the barbarian condition of their builders. There is here no lack of muscular energy. No people of native indolence could have erected these monuments, or have even conceived the idea of them. There is abundant ability to work displayed, but no great ability to think. The great roads of Peru are products of a practical mind. In regard to the remaining works, they were largely incited by religious thought. They yield us in massive walls and crude ornamentation the record of the highest imaginative out-reach and artistic power of the American mind. When we come to examine them we find that their main expression is that of hugeness. Their art is rudimentary, except in some few striking instances in the Maya architecture and statuary of Yucatan. There are indications of intellectual ability, but it remains in its undeveloped stage. Energy is not lacking, but it is mainly confined to the muscles, and but slightly vitalizes the mind.

We have evidences of similar conditions in the works of architecture remaining from the pre-Aryan age of Europe. The huge monoliths of Stonehenge, Avebury, and Carnac, and the Cyclopean walls of Greece and Italy (the latter possibly of Aryan formation), indicate a race or an era when muscle was in the ascendant and thought in embryo. The idea was the same as that indicated in the structures of Asia and America,—to astound future man with edifices that seem the work of giant builders. No indication of the loftiest conception of architectural art appears,—that of the simple combination of the ornamental with the practical, and the restriction of size to the demands of necessity and the requirements of graceful proportion. To astonish by mere hugeness is a conception of the undeveloped mind. Blind force can raise a mountain mass; only highly developed intellect can erect a Greek temple.

The Melanochroic division of the white race repeats in

its work the Mongolian characteristic of hugeness. Yet it indicates superior thought-powers, and has attained a much higher level of art. In the extraordinary architectural and artistic monuments of Egypt the power of sheer muscular vigor displayed is astounding. The world has never shown a greater degree of energy; but it is rather energy of the hands than of the mind. The rudimentary idea of vast size is the main expression of these works; and though they have sufficient artistic value to show a considerable mental unfoldment, yet hugeness of dimensions and the power of overcoming difficulties are their overruling characteristics. The old rulers of Egypt were eager to show the world of the future what labors they could perform; they were much less eager to show what thought they could embody.

And yet among the monuments of Egypt and those of the sister nations of Assyria and Babylonia we find ourselves in a circle of thought of far higher grade than that displayed by the Mongolian monuments. There is indicated a vigorous power of imagination and an artistic ability of no mean grade, while strong evidence appears that but for the restraint of conventionality and the distracting idea of hugeness, art would have attained a much higher level. The rudiment of the Greek temple appears in the architecture of Egypt and Assyria, and the former is a direct outgrowth from the latter in the hands of a people of superior intellectuality.

If the Negro is indolent both physically and mentally, the Mongolian energetic physically but undeveloped mentally, and the Melanochroi active physically and to some extent mentally, in the Aryan we find a highly vigorous and developed mental activity. Though by no means lacking in physical energy, the mind is the ruling agent in this race, muscular work is reduced to the lowest level consistent with the demands of the body and the intellect, and every effort is made to limit the quantity of work represented in a fixed quantity of product. Waste labor is a crime to the Aryan mind. Use is the guiding principle in all effort. It is to this ruling agency of the intellect over the energies of a muscular and active organism that we owe the superior quality, the restricted dimensions, and the vast quantity of Aryan labor products. In this work pure thought is far more strongly represented than pure labor.

In the two great intellectual Aryan peoples of the past, the Greek and the Hindu, the artistic products are strikingly in accordance with the character of their respective mentality. The work of the Hindu displays an imaginative exuberance, with a lack of reasoning control. In it we have rather the idea of vastness than of hugeness, a vague yet strong mental upreach, while a superfluity, almost a wildness, of ornament testifies to the unrestrained activity of the imagination. There is indicated no controlling idea of utility. The Hindus were almost devoid of practicality. Their architecture seems an embodiment of their philosophy, — daring, unrestrained, and unpractical throughout. In their older cave-temples, such as that at Elephanta, sheer labor is the strongest characteristic; but it is labor underlaid with a vigorous sense of art. In the extraordinary excavations at Ellora an exuberant imagination carries all before it, and we seem to gaze upon an epic poem in stone, rendered inartistic by its endless superfluity of ornament.

In Greek architecture and in all Greek art, on the contrary, are visible the evidences of a subdued imagination. In breadth and height of imaginative conception the Greek mind is in no sense inferior to the Hindu, but it is everywhere restrained by the habit of observation and by a sense of the logical fitness of things. The Hindu looked inward for his models, and built his temples to fit the conceptions of his imagination. The Greek looked outward, found his models in the lines and forms of the visible, and sought to bring his work into strict conformity with the grace, harmony, and moderation of external Nature. In this effort he attained a remarkable success. True art was born with him. All excess and exuberance disappears, the wings of the imagination are clipped, and its flights kept down to the level of the visible earth. The idea of the practical is everywhere combined with that of the ornamental. The subordination of the mind to the teachings of visible Nature is rigidly maintained. Greek art is the actual, reproduced in all its lines and proportions, and with a strictly faithful rendering that detracts from its value as a work of the intellect, while adding to it as a work of art.

The defect of Greek art lies in an excess of this restraint. It sins in one direction, as Hindu art does in the other. The wings of the imagination are too severely clipped. It is undoubtedly a high conception of art accurately to reproduce in marble the exact details and proportions of the human frame. But the Greek fixed his eyes so closely upon the body that he in a measure lost sight of its animating soul. This is not the highest conception of art. To imitate physical Nature exactly, was a great achievement; and this the Greek artist attained to a degree that can never be surpassed. But to reproduce the mind in the body, is a greater achievement; and in this direction Greek art made but the preliminary steps.

The great statues of Greece represent types, not individuals. They display the mental characteristics of fear, modesty, terror, dignity, and the like, in the gross, not in detail. Their works are like the combined photographs by which the general typical features of groups of men are now reproduced. The special and individual varieties of these characters are never represented. It is the same with Greek architecture. It contains the harmonies and proportions of physical Nature, but it is empty of the deep spiritual significance with which Nature is everywhere pervaded. It is a magnificent body, but it lacks the soul. The same would doubtless prove to be the case with Greek painting, had it been preserved. It is largely the case with Greek literature. Its characters are types of man more largely than they are individual men. Too strict devotion to the seen is the weak point in Greek thought. Its flight lies below the level of the unseen.

Modern Aryan art has taken a higher flight. While paying less attention to the body, it has paid more to the soul. In Gothic architecture the imagination displays a certain extravagance of manifestation; but in it there is embodied something of that profound and awe-inspiring spiritual significance of Nature which Greek art fails to manifest. Modern sculpture, while it does not attain to the Greek level of physical perfection, indicates a higher ideal of mentality. It represents the individual instead of the group, and seeks to reproduce human emotion in its special, instead of its general varieties of manifestation. But the true modern arts, those best suited for mental embodiment, are painting and music. Of these the former attained some ancient development; the latter is strictly modern as an art. It is mainly in these, and particularly in music, — the latest production of Aryan art, — that the soul shows through the thought, and that man has broken the crust of clay which envelops his inmost being, and animated the products of his art with the deep spiritual significance that everywhere underlies Nature. In the work of the modern artist, in fact, we seem to have found the true middle line between the opposite one-sidedness of Greek and Hindu art. In the former of these the visible too strongly controls; in the latter the invisible. In the one the logical, in the other the imaginative, faculty of the mind attains undue predominance. The modern artist seeks to make these extremes meet. He fails to rival the Greek in the physical perfection of his work mainly because his thought looks deeper than mere physical perfection; he fails to display the Hindu exuberance of fancy from the fact that he never loses sight of the physical. As a consequence, his work pursues the mid-channel between the logical and the imaginative, and reproduces Nature as it actually exists, — everywhere a body animated by a soul. It is the individual that appears in modern art, as it is the individual that rules in modern society. In ancient nations the individual was of secondary importance. The group was the national unit alike in the family, the village, the gens, the tribe, and the various subdivisions of the State. The individual was imperfectly recognized in society, and became as imperfectly recognized in art.

In respect to the art of the non-Aryan nations little need be said. It lay far, often immeasurably, below the level of Aryan art. What the art of Egypt might have attained if freed from the restraint of conventionalism, it is difficult to say. It would probably even then have ended where Greek art began, as we find to be the case with the less conventionalized art of Assyria. The art of the Americans was far more rudimentary. In one or two examples it approaches the character of Greek art, but as a rule it is rather grotesque than artistic. The same remark applies to the art of modern China. It belongs to the childhood of thought.

The world of science is almost completely an Aryan world. In this important field of thought the non-Aryan races of mankind stop at the threshold of discovery. Their most important work is in the formation of the calendar, to which strict necessity seems to have driven them. In this direction considerable progress was early attained. Each of the primitive civilizations measured the length of the year with close exactness, the Mexicans particularly so, their calendar being almost equally accurate with that of modern nations. This was a work of pure observation, and astronomical conditions seem strongly to have attracted the attention of early man. In fact the only extended series of scientific observations in the far past of which we are aware, is that of the Babylonians, in their close watch upon the movements of the stars and their study of eclipses. As to the accuracy and actual value of this work, we really know very little. Some similar observations were recorded by the Chinese. But nearly all the actual results of science which the Aryan has received from the exterior world consist in these few astronomical observations, — the partial settlement of the length of the year, its division into months and weeks, and the similar division of the day into its minor portions.

On this small foundation the Aryans have built an immense superstructure. Aryan science began with the Greeks, whose tendency to exact observation made them critically acquainted with many of the facts and conditions of Nature. Yet during all the early eras of Greek enlightenment the activity of the imagination prevented this habit of observation from producing valuable scientific results. It was devoted principally to the purposes of philosophy and art. It was necessary that able men, in whom logic was superior to imagination, should arise ere science could fairly begin. The first of these men we find in Thucydides, — a cool, practical thinker, who made history a science. The second of marked superiority was Aristotle, — the true founder of observational science, which had but a feeble existence before his day. His teacher, Plato, was a true Greek, with all the fervor of the Hellenic imagination. Aristotle was essentially a logical genius. An effort to bring himself into conformity with the prevailing conditions of Greek thought forced him into various lines of speculation; but the ruling tendency of his mind was toward incessant observation of facts for the accumulation of exact knowledge. There had been preceding Greek naturalists. Several noted physicians, particularly Hippocrates, had made medical investigations. Aristotle made use of the work of these men; but it is doubtful if it was of much extent or accuracy. To it he added a great accumulation of facts, while laying down the laws of logical thought, which he was the first to formulate, and to which little of value has been since added.

Any review of the subsequent history of science in the Aryan world is beyond our purpose. It is far too vast a subject to be even named at the conclusion of a chapter.

It will suffice to say that the Greek mind seized with avidity upon the new field of labor thus opened to it. It was native soil to Greek thought, although it yet lay fallow. The tendency of the Hellenic race to critical observation had for centuries been fitting them for the work of research into the facts of Nature; and had the Greek intellect remained in the ascendant there is no doubt that the schools of Alexandria would have been the focus of a great scientific development during the ancient era. As it was they performed a large amount of good work, and built a broad foundation for the future growth of this new product of the human understanding.

The Arabian empire served as the connecting-link between the thought of the ancient and modern world. We cannot exactly say the Arabians, for this broad empire clasped the thinkers of nearly all of civilized mankind within its mighty grasp. It handed down Greek philosophy and science to modern Europe, — the former with many additions but no improvements, the latter considerably advanced. The Arabian fancy played with Greek philosophy, but was incapable of developing it, or even of fully comprehending it. But observation and experiment needed no vigorous powers of the intellect, and in this direction many important discoveries were added by the Arabians to the science of the Greeks. As to the vast results of scientific observation of the modern Aryan world, nothing need here be said. The coffers of science are filled to bursting with their wealth of facts.

But science has by no means been confined to observation. The Aryan imagination has worked upon its store of facts as actively as of old it worked upon its store of fancies, and has yielded as abundant and far more valuable results. Nature is being rebuilt in the mind of man. One by one her laws and principles are being deduced from her observed conditions, and man is gaining an ever-widening and deepening knowledge of the realities of the universe in which he lives. And he is beginning to "know himself" in a far wider sense than was in the mind of the Grecian sage when he uttered this celebrated aphorism. The imagination of the past dealt largely with legend, with misconceptions of the universe, with half observations, and devised a long series of interesting but valueless fictions. The imagination of the present is dealing more and more with critically observed facts, and deducing from them the true philosophy of the universe, that of natural law, and of the unseen as logically demonstrable from the seen. This great field of intellectual labor belongs to the Aryans alone. The other races of mankind have not yet penetrated beyond its boundaries.

Modern Aryan civilization is made up of many more elements than those whose development we have hastily reviewed. One of the most marked of these is that of labor-saving machinery. This is somewhat strictly confined to modern times and to the Aryan nations. Beyond this limit it has never existed in other than its embryo state. Tools to aid hand-work have been devised, but the employment of other powers than the muscles of man to do the labor of the world is almost a new idea, scarcely a trace of it being discoverable beyond the borders of what we may denominate modern Arya. The immense progress made in the development of this idea is comparable with the unfoldment of science, and together they form the backbone of modern civilization. Knowledge of Nature, and industrial application of this knowledge, have given man a most vigorous hold upon the universe he inhabits; and in place of the slow, halting, and uncertain steps of progress in the past, he is now moving forward with a sure and solid tread, and down broad paths of development as firm and direct as were the great high-roads that led straight outward from Rome to every quarter of the civilized world.

The progress of commerce, of finance, and of inquiry into the underlying laws of social aggregation and political economy, has been no less great. Here, too, we must confine ourselves to the limits of the Aryan race, so far as modern activity is concerned. Commerce, however, had its origin at a very remote period of human history, and attained a marked development in Semitic lands before the Aryans had yet entered the circle of civilization. There is every reason to believe that the ancient Babylonians had a somewhat extensive sea and river commerce at a very remote epoch. They were succeeded by the Phœnicians, who displayed a boldness in daring the dangers of unknown seas that was never emulated by their successors, the Greeks. The overland commerce of the Phœnicians was also very extensive. Since the origin of Greek commerce, however, little activity has been shown in this direction by non-Aryan peoples, with the one exception of the Arabians, who carried on an extensive ocean commerce in their imperial era, and who to-day penetrate nearly every region of Africa in commercial enterprises. In this respect, also, modern China manifests some minor activity. Yet the Aryans are, and have been, the great commercial people of the earth, and have developed mercantile enterprise to an extraordinary degree. Commercial activity has been handed down in an interesting sequence from branch to branch of the Aryan race, the Greeks, the

Venetians, the Italians, the Portuguese, the Spanish, and the Dutch each flourishing for a period, and then giving way to a successor. To-day, however, commercial activity is becoming a common Aryan characteristic, and though England now holds the ascendency, her position is no longer one of assured supremacy. A century or two more will probably find every Aryan community aroused to active commercial enterprise, and no single nation will be able to claim dominion over the empire of trade. That any non Aryan nation will at an early period enter actively into competition in this struggle for the control of commerce, is questionable. The Japanese is the only one that now shows a strong disposition to avail itself of the advantages of Aryan progress, China yet hugging herself too closely in the cloak of her satisfied self-conceit to perceive that a new world has been created during her long slumber.

There is one further particular in which comparison may be made between the Aryan and the non-Aryan races of mankind, — that of moral development. In this direction, also, it can readily be shown that the Aryans have progressed beyond all their competitors. This, however, cannot be said in regard to the promulgation of the laws of morality, the great body of rules of conduct which have been developed for the private government of mankind. It is singular to find that no important code of morals can be traced to Aryan authorship, with the single exception of the Indian branch of the race. There we find the Buddhistic code, which is certainly one of remarkable character, but which has in very great measure lost its influence upon the Aryan race. Alike the morality and the philosophy of Buddhism have almost vanished from the land of their birth, and this religious system is now nearly confined to the Mongolian race, while its lofty code of moral observance has lost its value as a ruling force in the modern Buddhistic world.

A second great code of morals is that of Confucius, and constitutes essentially the whole of Confucianism. This religion of educated China consists simply of a series of moral rules, of a character capable of making a highly elevated race of the Chinese, had they any decided influence. They are studied abundantly, but only as a literary exercise. The moral condition of modern China indicates very clearly that the Confucian code is one of lip-service only. It has made but little impression upon the hearts of the people.

The third and highest of the three great codes of morals is of Semitic authorship, being the lofty doctrine of human conduct promulgated by Christ. So far as the mere rules of conduct embraced in it are concerned, it differs in no essential features from those already named. Its superior merit lies in its lack of appeal to the selfish instincts, and its broad human sympathy. Buddhism warns man to be virtuous if he would escape from earthly misery. Confucianism advises him to be virtuous if he would attain earthly happiness. Do good, that you may attain Nirvana. Do good to others if you wish others to do good to you. These are the dogmas of the two great non-Christian codes. Do good because it is your duty, is the Christ dogma. Sin defiles, virtue purifies, the soul. All men are brothers, and should regard one another with brotherly affection. "Love one another." This is the basic command of the code of Christ. And in this command we have the highest principle of human conduct, — a law of duty that is hampered by no conditions, and weakened by no promises.

It is singular that the creed of Christ has become the creed of the Aryan race alone. The Semites, even the Hebrews, of whose nation Christ was a scion, ignore his mission and his teachings. But throughout nearly the whole of the Aryan world it is the prevailing creed, and its code of morals is to-day observed in a higher degree than we find in the moral observance of the remainder of mankind. Elsewhere, indeed, there is abundance of private and local virtue, and rigidly strict observance of some laws of conduct, though others of equal value are greatly neglected. But nowhere else has human charity and the sense of human brotherhood attained the breadth they display in the Aryan world, and nowhere else can the feeling of sympathy with all mankind be said to exist. There is abundance of evil in the Aryan nations, but there is also abundance of good; and the minor sense of human duty which is elsewhere manifested is replaced here with a broad and lofty view that fairly stamps the Aryan as the great moral, as it is the great intellectual, race of mankind.

XII.

HISTORICAL MIGRATIONS.

WHEN history opens, it reveals to us the Aryan race in possession of a vast region of the eastern hemisphere, including some of its fairest and most fruitful portions. How long it had been engaged in attaining this expansion from its primitive contracted locality; what battles it had fought and what blood shed; what victories it had won and what defeats experienced, — on all this human annals are silent. But we may rest assured that many centuries of outrage, slaughter, misery, and brutality lie hidden in this prehistoric abyss. Millions of men were swept from the face of the earth, millions more deprived of their possessions, and even of their religions and languages, millions incorporated into the Aryan tribes, during this expansion of primitive Arya. The relations of human races, which had perhaps remained practically undisturbed for many thousands of years, were largely changed by this vigorous irruption of the most energetic family of mankind. It was as if an earthquake had rent the soil of human society, broken up all its ancient strata, and thrown mankind into new and confused relations, burying the old lines of demarcation too deeply to be ever discovered.

The Aryan migration displays the marks of a high vigor for so barbaric an age, and was probably the most energetic of all the prehistoric movements of mankind. It met with no check in Europe except in the frozen regions of the extreme North, and there it was Nature, not man, that brought it to rest. Such also was probably the case in northern Asia. The deserts and the mountain-ranges there became its boundaries. China lay safe behind her almost impassable desert and mountain borders. In the south of Asia only the Semites held their own. They offered as outposts the warlike tribes and nations of Syria and Assyria. Possibly an era of hostility may have here existed; but if so it has left no record, and there is nothing to show that the Aryans ever broke through this wall of defence. But the remainder of southern Asia fell into their hands, with the exception of southern India with its dense millions of aborigines, and the distant region of Indo-China, on whose borders the Aryan migration spent its force.

Such is the extension of the Aryan world with which history opens. It embraced all Europe, with the exception of some minor outlying portions and probably a considerable region in northern Russia. In Asia it included Asia Minor and the Caucasus, Armenia, Media, Persia, and India, with the intermediate Bactrian region. These formed the limits of the primitive Aryan outpush, and it is remarkable that it failed to pass beyond these borders, with the exception of a temporary southward expansion, for two or three thousand years. It made some external conquests; but they were all lost again, and at the opening of the sixteenth century the Aryan race was in possession of no lands that it had not occupied at the beginning of the historical period.

This is a striking circumstance, and calls for some inquiry as to its cause. What was the influence that placed this long check upon the Aryan outflow? The acting influences, in fact, were several, which may be briefly named. A chief one was the almost insuperable obstacle to further expansion. Many of the boundaries of the new Aryan world were oceanic, and the art of navigation was as yet almost unknown to the Aryan race. Other boundaries were desert plains that offered no attraction to an agricultural people. The purely pastoral and nomadic days of the race were long since past. In the East the boundary was formed by the vast multitudes of Indian aborigines, who fiercely fought for their homes and made the Hindu advance a very gradual process. In the South warlike Assyria formed the boundary, and the Semitic world sternly held its own.

As Aryan civilization progressed, the great prizes of ambition were mainly included within the borders of the Aryan world. There is no evidence of a loss of the original migratory energy; yet it was no longer an energy of general expansion, but of the expansion of the separate branches of the race. The Aryan peoples made each other their prey, and the outside world was safe from their incursions. The only alluring region of this non-Aryan world was that of the Semitic nations and of Egypt. This fell at length before Aryan vigor, and became successively the prey of Persia, Greece, and Rome. And the thriving settlements which the Phœnicians had established in northern Africa fell before the arms of Rome. Such was the only extension of the borders of the Aryan world which history reveals, and this extension was but a temporary one. After a thousand years of occupancy the hold of the Aryans upon the Semitic and Hamitic regions was broken, and the invading race was once more confined within its old domain.

It is not necessary to repeat in detail the historic movements of the Aryans of ancient times. These are too well known to need extended description. They began with the rebellion of the Medes against Assyrian rule, and with the subsequent rapid growth of the Persian empire, which overran Assyria, Syria, and Egypt. At a later date the Greeks made their great historical expansion, and under Alexander gained lordship over the civilized Aryan world. Still later the Romans established a yet wider empire, and the world of civilization was divided between Rome and Persia. The *finale* of these movements was the irruption of the Teutons upon the Roman empire, which buried all the higher civilization under a flood of barbarism.

Thus for about a thousand years the great battle-field of the world had been confined mainly within Aryan limits, and the other races of mankind had remained cowed spectators, or to some extent helpless victims, of this bull-dog strife for empire. The contest ended with a marked decline in civilization and a temporary loss of that industrial and political development which had resulted from many centuries of physical and mental labor. The Aryan race had completed its first cycle, and swung down again into comparative barbarism, under the onslaught of its most barbarous section, and as a natural result of its devastating and unceasing wars.

And now a remarkable phase in the history of human events appeared. The energy of the ancient Aryan world seemed to have spent its force. That of the non-Aryan world suddenly rose into an extraordinary display of vigor. The Aryan expansion not only ceased, but a reverse movement took place. Everywhere we find its borders contracting under a fierce and vigorous onslaught from the

Mongolian and Semitic tribes. This phase of the migratory cycle we may run over as rapidly as we did that of the expanding phase.

The first marked historical movement in this migratory series was that of the Huns, who overran Slavonic and pushed far into Teutonic Europe, and under the fierce Attila threatened to place a Hunnish dynasty on the throne of imperial Rome. The next striking movement was the Arabian, which drove back the wave of Aryan conquest from the Semitic region, from Egypt, and from northern Africa, and brought Persia and Spain under Arabian domination. The third was that of the Turks, who replaced the Arabian rulers of Persia, conquered Asia Minor, and finally captured Constantinople and the Eastern Empire, extending their dominion far into Europe and over the Mediterranean islands. The fourth was that of the Mongols, under Genghiz Khan and Timur, which placed a Mongol dynasty on the throne of India and made the greater part of Russia a Mongol realm. We need not mention the minor invasions, of temporary effect, which broke like fierce billows on the shores of the Aryan world and flowed back, leaving ruin and disorder behind them. It will suffice to describe the contraction of the borders of the Aryan region which succeeded this fierce outbreak of the desert hordes upon the civilized world.

All the historical acquisitions of the Aryans were torn from their hands. The Semitic region became divided between the Turks and the Arabians. Egypt and northern Africa were rent from the Aryan world. In the East, Persia, India, and the intermediate provinces, though with no decrease in their Aryan populations, lay under Mongol rule. In the West, Spain had become an Arabian kingdom.

A Hungarian nation in central Europe was left to mark the onslaught of the Hunnish tribes. In eastern Europe, the Tartars occupied Russia in force, and held dominion over the greater part of that empire. Farther south, the Turks were in full possession of Asia Minor and Armenia, held the region of ancient Greece and Macedonia, and extended their barbaric rule far toward the centre of Europe. The contraction of the ancient Aryan region had been extreme. As a dominant race they held scarce half their old dominions, while in many regions they had been driven out or destroyed, and replaced by peoples of alien blood.

Such was the condition of Europe at the close of the Middle Ages. The first cycle of human history had become completed, the expansion of the Aryans had been succeeded by a severe contraction, the growth of ancient civilization had been followed by a partial relapse into barbarism, human progress had moved through a grand curve, and returned far back toward its starting-point. Such was the stage from which the more recent history of mankind took its rise.

It may be said that of the energy of the Aryans and the non-Aryans the former has proved persistent, the latter spasmodic. No sooner was the condition of affairs above mentioned established than the unceasing pressure of Aryan energy again began to tell, and a new process of Aryan expansion to set in. And this process has been continued with unceasing vigor till the present day. The Aryans of Spain began, from a mountain corner, to exert a warlike pressure upon the Arabian conquerors of their land. Step by step the Arabs were driven back, until they were finally expelled to the African shores. Simultaneously a vigorous effort was made to wrest Syria from its Arab lords. All

Europe broke into a migratory fever, and the Crusades threw their millions upon that revered land. But all in vain. The grasp of the Moslem was as yet too firm to be loosened by all the crusading strength of Europe.

At a later date the Mongol hold was slowly broken in Russia, and the Slavonic Aryans regained control of their ancient realm, while the invasion of the Turks was checked, and a reverse movement begun which has continued to the present day. As for the Magyars of Hungary, their realm has been partly reconquered by Aryan colonists, its civilization and government are strictly Aryan, and the Mongolian characteristics of the predominant race have been to a considerable extent lost. Europe has been reoccupied by the Aryans, with the exception of a few Turks who are left upon its borders by sufferance, and the Mongoloids of the frozen North. In Asia the Aryan spirit has declared itself less vigorously; yet Persia, Afghanistan, and India have declined little if at all in the percentage of their Aryan populations, while Aryan dominance has replaced the Mongol rule in India. As for the Aryan physical type, it seems to be killing out the type of the Mongolian in all regions exposed to its influence. Thus the Osmanli Turks have gained in great measure the European physical organization, this applying even to the peasantry, whose religious and race prejudices must have prevented much intermarriage with the Aryans. It looks, in this instance, like an effect of climate, physical surroundings, and life-habits similar to that which, as we have conjectured, caused the original evolution of the Aryan race. The same influences may have had much to do with the loss of Mongolian characteristics in the Magyars of Hungary.

But the Aryans have been by no means contented with this slow and as yet but partially completed recovery of their ancient realm. Only the mutual jealousy of the nations of Europe permits aliens yet to occupy any portion of this soil, and it is plainly apparent that the complete restoration of Aryan government over all its ancient dominions is a mere question of time. But the slow steps of this internal movement have been accompanied by an external one of vast magnitude. After its long rest the Aryan race has again become actively migratory, an expansive movement of great energy has set in, and the promise is that ere it ends nearly the whole of the habitable earth will be under Aryan rule, infused with Aryan civilization, and largely peopled with Aryan inhabitants.

It is the control of the empire of the ocean that has been the moving force in this new migration. The former one was checked, as we have said, upon the ocean border. Navigation had not yet become an Aryan art. But the rise of ocean commerce gave opportunity for a new outpush of no less vigor than that of old. When once the European navigators dared to break loose from sight of land and brave the dangers of unknown seas, a new chapter in the history of mankind began. The ships of Europe touched the American shores, and with phenomenal rapidity the invaders took possession of this new-discovered continent. Not four centuries have passed, and yet America, from its northern to its southern extremities, is crowded with men of Aryan blood, and the aborigines have in great measure vanished before the ruthless footstep of conquest.

In the East the activity of Aryan migration has had more difficulties to contend with, yet its energy has been no less declared. The island continent of Australia has become an outlying section of the Aryan dominions, and in many of the fertile islands of the Pacific the aborigines are rapidly vanishing before the fatal vision of the European face. The non-Aryan rulers of India have been driven out, and England has succeeded to the dominion of this ancient realm. And finally the "dark continent" of Africa is being penetrated at a hundred points by the foot of the invader, and is already the seat of several Aryan states.

Side by side with this oceanic migration has been a no less active and important expansion by land. The Slavonic Aryans of Russia had no sooner fairly driven out their Tartar conquerors and acquired a stable government than they resumed their ancient migratory expansion and began to press their way into that vast region of northern and central Asia upon whose borders the ancient Aryan advance had paused. Siberia fell before their arms, and this great but frozen region was added to their empire. More recently they have taken possession of the western steppes, seized a considerable region of Chinese Mongolia, and forced their way deeply into Turkestan. All western Asia to the borders of China, Afghanistan, and Persia is to-day a Russian province, and still the march of conquest goes on. Of the regions of the ancient non-Aryan migratory activity none, with the exception of Arabia and Chinese Mongolia, is free from the Aryan grasp or the preventive influence of Aryan control. The barbarian outbreaks of the past can never be repeated.

In regard to this modern migratory activity some further remarks may be made. It is in a great measure a commercial one, and has been very closely governed in its movements by those of commerce. It had its origin in the Phœnician trading-stations, and subsequently in the Greek colonies. It passed from branch to branch of the Aryan peoples in strict accordance with the shiftings of commerce. At the period of the discovery of America there was a very general commercial activity in the Atlantic nations of Europe, and all of these simultaneously took part in the struggle for territory that followed. Portugal, Spain, France, Holland, and England each claimed a share in the rich prize. At a later date, however, England rose to unquestioned supremacy in the commercial world, and this was accompanied by a similar rise to supremacy in colonizing efforts. The England of to-day is extended until it has its outlying members in almost every region of the habitable earth. The other Aryan peoples, on the contrary, with the exception of Russia, have lost in great measure their national migratory activity, as they have lost their commercial enterprise. The Celts and Germans still migrate largely as individuals, but this migration mainly goes to feed colonies of English origin and to add to the English-speaking populations of the earth. The very recent colonizing movements of Germany are acts of the Government, and it remains to be seen if they will be supported by the people. The same may be said of the colonial enterprises of France. They are Governmental enterprises only, while the people are among the least migratory in spirit of any European nation. Only in England, of all the commercial nations of Europe, are the people and the Government moving hand in hand.

Thus the Aryan migration has to-day reached a highly interesting stage. The boundary lines which restrained it several thousand years ago and which remained its limits until within recent times, have been overleaped, and a new migration, with all the energy of the old one, is in process of completion. This migratory movement is at present largely confined to two of the Aryan peoples, — the English and the Russian. The former has broken through the ocean barrier; the latter through the desert barrier, — the two limits to the ancient migration. The English movement is entirely oceanic, the Russian entirely terrestrial. The English represents the modern commercial migration; the Russian is a survival of the primitive agricultural migration. These two peoples form the vanguard of the Aryan race in its double march to gain the empire of the earth. By a strange coincidence their movements converge upon one region, — that of India, one of the great prizes of commerce and war in all the historic ages of mankind. On the borders of this land the two waves of migration have nearly met, and the lords of the land and the sea threaten to join in battle for its mastery. Aryan is again face to face with Aryan as in the era of the past, and, as then, the migratory march may end in a fierce strife of these ancient cousins for a lion's share of the spoils.

The Aryan outposts of to-day are being pushed forward so rapidly that they cannot be very definitely named. The whole of the great continent of America has become an Aryan region, with the exception of the inaccessible forests of central Brazil and some few minor localities. In the eastern seas the great island of Australia has become Aryan ground to the inner limit of its fertile land. In most of the rich islands of the Pacific the Aryan grasp has been firmly laid upon the coast-regions, though the aborigines as a rule hold their own internally. The vegetable wealth of these fertile islands has become the prize of Aryan commerce. In Asia one of the ancient Aryan lands, the kingdom of Persia, is under Mongolian rule, though its population continues largely of Aryan blood. But in return the greater portion of the old Mongolian territory has fallen under Aryan dominion, and the outposts of European rule have been pushed across Asia to the Pacific in the north, and to the western borders of China in the central region. Again, in the southeast, in that remote region which stayed the march of the ancient Aryans, the modern Aryans are slowly pushing their way. England years ago laid her hand on the western coast-lands and occupied the maritime region of Burmah, while she has recently seized on the whole of that kingdom. France has taken as firm a hold on the eastern coast, over which she exerts a controlling influence. Siam, the remaining independent region of Indo-China, will probably yet fall under the rule of these enterprising invaders.

Africa tells a somewhat similar story. France has regained from the Mohammedan rule a large section of the old Roman region in northern Africa. England has become the virtual lord in Egypt, and may eventually become the acknowledged lord. Southern Africa, for a long distance northward from the Cape, has become English and Dutch territory. Portugal holds large districts on both the eastern and the western coasts. Of the remaining coast-lands, all the western border and a considerable portion of the eastern are claimed by European nationalities, while in the region of the Congo a strong inward movement is on foot, and the International Association lays claim to an immense territory in Central Africa, — a region with a population of perhaps forty millions, who do not dream that they have gained new lords on paper. Such is the border-land, actual and claimed, of modern Arya, — the result of four centuries of commercial and colonial enterprise. The Aryan region of old has been much more than doubled by this new movement. The hold is yet to some extent simply the grasp of an army or of a document. But the colonist is advancing in the rear of the army, and the merchant in the rear of the document; and the story of Aryan enterprise is but half told.

If now we seek to review what the other races of mankind have done, in rivalry with this energetic movement, a few words will suffice to tell the tale. The alien outflow is confined to three peoples alone. The first of these is the Chinese, some portion of whose crowding millions are forced to seek other homes afar, and whose strongly practical disposition has produced a degree of commercial enterprise. Yet the results of this movement have been as yet of secondary importance. It has made itself felt in some regions of the Pacific, and to a minor extent in America. Yet it can never attain a vigor comparable to the Aryan while Chinese civilization and Chinese ideas remain in their present state. The Chinaman is not yet cosmopolitan like the Aryan; the world is not his home; and wherever he goes he dreams of laying his bones to rest in Chinese soil. While such ideas persist, the Aryans need fear no powerful competition from this ancient realm. As for the neighboring Japanese, they have so far shown no disposition to wander. They are in no sense a migratory people.

The second non-Aryan migratory people is the Arabian. The migratory spirit which has in all historic times affected the Semites has by no means died out; and while Europe is grasping the African shores, the Arabs are penetrating every portion of the interior of that continent. But their movements are commercial only, not colonial. The sole political grasp of Arabia on African soil is in the region of Zanzibar. Elsewhere their political dominion is but that of the wandering tribe. The Arabs of to-day are not in the state of civilization requisite to active colonization, while there is no pressure of numbers in the home region to enforce a border outgrowth. Thus there can be said to be no combined Arabian competition with the Aryans for the political possession of Africa. The empire-forming enterprise of the Arabians of old has apparently died out; and while they retain all their ancient commercial activity, they manifest no inclination to gain political control of African soil.

The third migration referred to comes from Africa itself. It no longer exists, but has had the unfortunate effect of very considerably extending the area of the Negro race, — the least-developed section of the human family. This migration has been solely an involuntary and unnatural one. It is not the outcome of enterprise among the migrants, but of the enslaving activity of the Aryans, and has resulted in widely extending the limits and increasing the numbers of the most unenterprising and unintellectual of human races. The migration of Africans to the shores of America has proved a highly undesirable result of Aryan enterprise, and has produced a rapidly increasing population of American Negroes, who cannot but remain an awkward problem for the civilization of the future. This people has the unlucky characteristic of prolific increase, and the unsealing of the continent of Africa by the slave-dealers has proved like the unsealing of the magic jar brought up in his net by the Arabian fisherman. A living cloud has issued, which cannot be replaced in its former space, and the sealed-up dwarf has been permitted to expand to the stature of the released giant. This enforced outpour of the African race is one of the several unfortunate results of the over-greed of Aryan colonists. It has proved far the most unfortunate feature of modern migratory activity by its extension of the domain of low intellectuality upon the earth.

We may close with one further consideration, — that of the comparative good and evil resulting from this modern Aryan outgrowth. That it has been conducted brutally, no one would think of denying. The laws of morality and of natural right have been abrogated in dealing with alien races; and had these been wild beasts instead of men, they in many cases could not have been more cruelly treated or rapidly annihilated. Yet if we could strictly compare the good and evil produced, there can be no question that the former would, so far as man as a whole is concerned, far outweigh its opposite.

What are the actual facts concerning the suffering which the aborigines of the earth have endured from Aryan hands, and the change for the worse in their condition produced by Aryan occupation? The treatment of the American Indian is usually considered as a flagrant example of injury to the aborigines. Yet it cannot be justly said that the Indians of the United States have been at any time visited with more suffering, and made the subjects of greater outrage, during the Aryan occupation, than they were ordinarily exposed to before that occupation. The preceding period was one of incessant war, outrage, slaughter, and torture of prisoners. Security nowhere existed, and it was impossible for any civilizing progress to take place. The wars which the Indians waged with the Europeans were but a continuation of those they had always previously waged. The slaughter of Indians was in no sense increased, while there was produced a mitigation of the more revolting features of Indian conflict. And the Aryan wars with the Indians were waged in the interests of peace. They have steadily decreased in violence and frequency, and an increasing justice and security in the conditions of Indian life have replaced the old rule of injustice and insecurity, which but for the European colonization would still have continued. It may safely be declared, then, that the Indians have been benefited far more than they have been injured by the Aryan conquest, and that to-day they exist in a far higher state of security, comfort, and happiness than they would have attained if that conquest had not been made.

Similar remarks can be applied to the Aryan conquests in every region, with the one exception of Spanish America. Here two civilized empires were overturned by colonists whose civilization was, in certain respects, of a lower grade, and millions of people were reduced from a state of plenty, and comparative freedom and happiness, to one of want, slavery, and misery. And yet, so far as the actual progress of civilization is concerned, the general interests of mankind have not suffered by this outrage. A civilization of a higher grade has succeeded the imperfect conditions of the Aztec and Peruvian States, and the mass of the human inhabitants of these regions are in a superior condition to-day than they would have been but for the Aryan conquest. The low conditions of Indian have been replaced by the high conditions of European civilization.

This Spanish region, however, is the one black spot in the history of modern migration. Elsewhere the good has far surpassed the evil. No one can for a moment hold that the Africans or the Australians are the worse off for the Aryan settlements upon their soil. Nor can it be maintained that an extension of these settlements will work any actual harm to the aborigines. At present they are in a debased condition, and are subject to constant outrage and injustice from their rulers or from hostile bands. The influence of Europeans is steadily in the interest of peace, security, and prosperity; and fiercely as they have been often opposed by natives of the countries colonized, yet as a rule these natives have been fighting against their own advantage. Wherever the Aryan race has become definitely established, and peaceful conditions succeeded, the condition of the natives has been improved, the wealth of their country developed, all the needs of a comfortable life increased, peace has succeeded to war, security to outrage, and the happiness of mankind has steadily augmented.

The true effect of Aryan migration has been the extension of the realm of modern civilization, of Christian ethics, of stable and just political conditions; of active industry, peaceful relations, and security in the possession of property; of human liberty and intellectual unfoldment; of commerce and developed agriculture; of railroads, telegraphs, books, tools, abundance of food, lofty thoughts, and high impulses; and of the noblest standard and most unfolded practice of morality and human sympathy the world has yet attained. We can scarcely name in comparison with this great benefit the small increase of evil, the degree of human suffering which can be attributed to the Aryans alone, in excess of that which would have existed without them. As a whole it must be admitted that the Aryan migration has acted and is acting for the best interests of all mankind; and it cannot consistently be deprecated for the minor amount of evil it has originated.

XIII.

THE FUTURE STATUS OF HUMAN RACES.

ONE important effect of the long process of human evolution which we have considered in the preceding pages has been such a mingling of the races of mankind as in considerable measure to blur the lines of race-distinction. This mingling, which began in prehistoric times, has proceeded with enhanced rapidity during the historic period, — that of active migration and of decreasing devastation. The movements of savage races and of races in the lower stages of barbarism are apt to be annihilating ones. Of this we have historic instances in the wars of the American Indians, of the Mongolian nomads, and even of the Anglo-Saxon conquerors of England. The captive must have some value to the conqueror ere he will be permitted to live, and the practice of slavery produced the first great amelioration of human brutality. The captors ceased to burn or otherwise slaughter their captives when they discovered that a slave was of more value than a corpse; and the class of conquered subjects who had been previously massacred were now set to work. In modern times a second step forward has been taken. The captive is no longer made the personal slave, but merely the political subject of the captor, and the ancient feeling of hostility to the non-combatant is rapidly dying out. Migratory peoples no longer make a desert for the growth of their colonies, but simply establish their laws and introduce their customs in all newly occupied regions, and mingle freely with their new subjects.

The result of this is necessarily a considerable obliteration of race-distinctions. Such an obliteration has been visibly going on since the early days of history, while many traces of its prehistoric activity yet exist. We have already dwelt upon the probable partial mingling of the Xanthochroic and Melanochroic races in ancient Arya. This was succeeded by a considerable fusion of the migrating Aryans with the aborigines of conquered provinces. The almost pure Xanthochroi of the original Celtic migration appear to have so thoroughly mingled with a superabundant population of European aborigines as nearly to lose their race-characters, and to suffer marked changes in their mental constitution. In Hindustan a similar mingling, though probably a less complete one, took place. Religious antipathy here acted as a check of growing intensity to race-amalgamation. An active race-mingling appears to have taken place in Germany and Russia. Scandinavia remained the only home of people of pure Xanthochroic blood. The probability is, as we have already suggested, that the southern Xanthochroi had mingled with the Melanochroi at a very early period, but that the infusion of alien blood was much less decided in the northern section of the race, and that the northern Aryan migrants were nearly pure Xanthochroi. Such seems to be the case from the fact that their most northerly portion is yet of pure blood, and that this was the condition of the Celts and Teutons of early history. The main mingling with the Semitic Melanochroi was probably that of the southern branches, who may have been, from a very remote period, in direct contact with the Semites. The mingling of the other Aryan branches with alien races seems to have mainly taken place after the era of their migration.

As we have seen in the last section, however, the completion of the original Aryan migration was succeeded by a long period in which the main Aryan movements were confined to Aryan lands. There was a very considerable mingling of blood between the different branches of the Aryans, but the amalgamation with alien races was greatly reduced. Almost no mixture with the Mongolians took place. To the south, however, there was more mingling, and the Semites and Hamites must have received a strong infusion of Aryan blood. This period was followed by that of the Arabian and the Mongolian migrations and conquests, and a very considerable new blood-mixture occurred upon Aryan soil. In Russia and in the Aryan districts of Asia this must have added very considerably to the obliteration of race-lines in those regions. Yet with all the long-continued amalgamations we have here considered, it is remarkable with what vigor the Aryan holds his own. His vital energy everywhere bears him up against alien influences. The main change produced in his race-characteristics is that of color. He varies greatly from fair to dark, but his special physiognomy has been nowhere obliterated. The Mongolian type of face has nowhere driven out the Aryan, but, on the contrary, shows a disposition to vanish whenever the two races come into contact. In like manner the Aryan language and the Aryan mentality have held their own against all opposing influences. This is the case in Persia and India, which have been the seat of the fiercest Mongolian inroads, while the Mongolian invaders of Turkey have lost in great measure the physical characters of their race, partly by intermarriage, but equally where no apparent intermarriage has taken place.

The more recent era of Aryan migration has not been an annihilating one in the ancient sense. Yet it has had a very marked annihilating effect in a modern sense. The migrants to America, for instance, have not greatly reduced the numbers of the aborigines by the sword; but they have largely destroyed them by the contact of civilization. They have brought with them diseases, habits, and vices to which civilization has become acclimated, but which have flowed like destroying angels over the barbarian lands. Rum and the small-pox have killed far more than the sword, while the plough has ruined the harvest of the arrow. In Spanish America hard work and brutality have had a similar effect. The race-mingling between the Aryan colonists and the Indians has been comparatively slight. There has been simply an industrial struggle for existence, and the Indian, from his non-adaptation to those new life-conditions, has in great measure vanished from his ancient localities. His place has been filled by a less desirable element, — that of the African, whose millions perhaps fully replace all the vanished aborigines of America. If so, the non-Aryan inhabitants of America are as numerous as ever, while they have been lowered in type both physically and mentally by this unfortunate change.

As to the future of human races in America, no satisfactory decision can be reached. The problem is a highly complex one. America is a grand storehouse of nations, the reservoir of the overflow from the Old World. Between the Aryan sections of this migration a very free mingling takes place, and there is arising an American race-type of well-marked character. There has also been considerable mingling of Aryan with Indian, particularly in Spanish America. As the Indians become civilized and agricultural in habits, it is probable that this amalgamation will go on at an increased rate, and it is quite possible that the Indians may finally disappear as a distinct race, swallowed up by the teeming millions of Aryan colonists. If they hold their own, it will be in the tropical regions of South America, where the conditions of Nature are opposed to the progress of civilization. Yet we can scarcely doubt that civilization will yet conquer even the Brazilian forests, and that the debased aborigines of that region will vanish before it.

The one perplexing problem of America is the Negro. Between him and the white the race-antipathy seems too strong for any great degree of amalgamation ever to take place, while the mulatto has the weakness and infertility of a hybrid. In tropical America, indeed, there is a quite free mingling of whites, Indians, and Negroes; but the result of this amalgamation is a class that greatly lacks staying qualities. The American Negro has marked persistence, while there is little promise that he can be raised to the level of Aryan energy and intellect. Mentally his only strong development is in the emotional direction, — the most primitive phase of mental unfoldment. Yet he is increasing in numbers with a discouraging rapidity. In this, however, there seems no threat to Aryan domination. The negro is normally peaceful and submissive. His lack of enterprise and of mental activity must keep him so. Education with him soon reaches its limit. It is capable of increasing the perceptive, but not of strongly awakening the reflective, faculties. The Negro will remain the worker. There is nothing to show that he will, at least for a long period to come, advance to the rank of the thinker. Of the two great modern divisions of civilized mankind, the workers and the thinkers, the Negro belongs by nature to the former class. He will probably long continue distinctly separate from the Aryans as a race, — a well-marked laboring caste among the non-differentiated whites of America.

As to the future of the continent of Africa, it may pass through conditions somewhat similar to those that have taken place in America; but these changes will be attended with less barbarity, since the moral status of the white race has very considerably advanced during the past four centuries. The wave of Aryan migration has as yet but begun to break upon African soil. Only in the far South has it pressed to any extent inward. But an inward pressure has now fairly set in, and it perhaps not cease until Africa has come completely under Aryan rule, and is very largely peopled by Aryan inhabitants. The Aryan settlements in the South promise to become paralleled by Aryan settlements in the North. Algiers is now a French province, Tunis is on the road to the same condition, and Morocco is threatened both by France and Spain, while Egypt is under English control. The march of events cannot go backward. There is very little reason to doubt that the whole region of northern Africa will eventually come under Aryan influence and become the seat of a growing Aryan population. And here a decided race-mingling will very probably take place in the future, as between the two sub-types of the Caucasian people in the far past.

Central Africa is being invaded by both these sub-types. Of these invasions the Melanochroic is to a considerable extent an amalgamating one. Between Arab and Berber and Negro, probably of close original race-affinity, there seems very little blood-antipathy; and Africa is full of sub-types of man, produced quite probably by a free mingling of the black with the Melanochroic race. How long this mingling has been going on, it is impossible to decide, and it is equally impossible to conjecture to what varied race-combinations in the far past the present inhabitants of Africa are due. But it is very evident that the future dealings of the Aryans with the Africans will not be conducted to any important extent with the race-counterparts of the American Negro. The American slaves were principally brought from nearly the only region of Africa inhabited by the typical Negro, and they thus represent the least-developed people of that continent. The majority of the African people are by no means lacking in energy and warlike vigor, nor in the elements of intelligence. Many of them seem to stand midway in these characteristics between the pure Negro of the western tropics and the Arabs and Berbers of the North. And the vanguard of Aryan migration may meet as hostile and resolute a resistance as that experienced from the American Indians.

The whole western coast of Africa, and to some extent the eastern, is at present dotted with Aryan colonies. None of these penetrate far inward, the unhealthfulness of the climate more than the opposition of the Negro checking their advance. But the key to the centre of the continent has been found in a great navigable river, the Congo, whose affluents spread far their liquid fingers through that fertile unknown land. In this line Aryan migration has fairly begun its inward march. It will meet with hostile tribes. Wars will take place. Forcible seizure and extinguishment of African governments will follow. Aryan control will be established over African populations. Many of the Africans will vanish before the Aryan weapons of rifle and whiskey-bottle. All this may be looked for as an almost inevitable consequence of the discovery that the Congo offers a new and valuable channel of commerce. The railroad past the rapids, and the steamboat on the river, cannot fail to subdue Central Africa, — far more quickly, perhaps, than the plough subdued America. Eventually this inward movement may meet with a northward movement from the South-African settlements. Nor is it possible at present to decide what may be the final outcome of English wars in the Soudan and in Abyssinia, and of French settlements in Algeria. For years past the Aryan influence in these regions has been steadily on the increase, and it may eventually make its way deeply into Africa from these directions toward the Aryan vanguard pressing inward from the West. A railroad is already pushing southward in Algeria, which may eventually cross the Sahara and reach the long-hidden city of Timbuctoo, toward which a railroad is also advancing from the South. As yet little more has been done than was accomplished by the Aryans in America during the sixteenth century. But there is every reason to believe, from what we know of the Aryan and the African character, that the final result will be the same. Africa will become a new empire of the Aryans. But the position of the migrants will be rather that of a ruling than of an inhabiting race. The condition of the Africans is markedly different from that of the Indians. They are much less warlike, and much more agricultural. They will undoubtedly remain upon the soil as its cultivators, while the *rôle* of the Aryans will be that of merchants, rulers, and artisans, in accordance with their position as the thinking and dominant minority. In fact there is some reason to believe that the march of events in the future will bring the African and the American continents into conditions of some degree of similarity. Through all the warmer regions of America the Negroes are increasing with great rapidity. They exist, and long may exist, as a working caste under Aryan dominance. Some similar relation of Aryans and Africans is not unlikely to arise on African soil, and the final relation of races in the warmer tropics of both hemispheres may be that here indicated, — a large population of African agricultural laborers, adapted by their physical nature to a tropical climate, and a smaller population of Aryan merchants, artisans, and rulers, mainly escaping the deleterious influence of tropical climates by city residence. In the higher and more healthful tropics and the semi-tropics the Aryan population must approach in numbers that of the tropically adapted race; and it must retain a great numerical excess, as now, in the temperate regions, to whose climate the Aryan is physically adapted.

That a race-mingling will take place between these two widely distinct types of man seems now extremely improbable. For a very long period to come it is certain that the physical and mental antipathy which now exists will be in no important degree overcome, and for many centuries in the future the demarcation may remain as strongly declared as now. What the final race-relation will be it is impossible to predict. There is no strong antipathy between the native races of the temperate zones of the earth, the Aryan, Indian, Mongolian, and Melanochroic; and these may mingle in an increasing ratio until their race-distinctions in great measure disappear. In such a case the only marked race-demarcation remaining will be that of white and black, respectively the man of the temperate and the man of the tropical climates of the earth. But the Indians of America and the Melanochroi of Africa have but little race-antipathy to the Negro, and their offspring is of a higher type than that of the Aryan and the Negro. It is possible, therefore, that the pure black may eventually vanish in an intermediate race, as is already so largely the case in Africa.

In the island region of the Pacific it is highly probable that the Aryan dominion, which is now firmly established in every island of any marked agricultural value, will grow more and more decided, and that the aborigines, or their Malayan successors, will eventually fall generally under Aryan rule. The lower aborigines will very probably vanish. They lie too far below the level of civilized conditions to survive the contact with civilization; and only those of declared agricultural habits, and the active Malays, are likely to remain as subjects of the growing Aryan rule.

There remains the probable future of the Aryans in Asia to pass in review. Here we find almost everywhere the same determined Aryan advance. During the last century the Aryan empire in Asia has been very greatly increased in dimensions. Nearly every trace of non-Aryan rule has been swept from India. Burmah promises to become an English province. The eastern coast of Indo-China is rapidly becoming a French one. If we may judge from past history, Siam, the only province of that region which yet fully retains its independence, will eventually fall under Aryan control. Persia, after being successively overrun by Arab, Turk, and Mongol, is to-day mainly Aryan in the race-characteristics of its civilized inhabitants. The Afghans and Belooches are principally Aryan. The whole of Asia to the north of the regions here mentioned, with the exception of the Chinese empire, is to-day under Russian rule, and becoming rapidly overrun by Russian merchants and colonists. That a very general race-mingling will eventually take place throughout this wide region is probable. The distinctive Mongolian features and mental conditions will become modified, and there can be little doubt that the Slavonic type of language will gradually crush out the less-cultured tongues of the region named.

In southwestern Asia there remain the Semites of the desert region and the Turks of Syria and Asia Minor. The latter would to-day be under Russian rule but for the jealousy of Europe. As a race they are becoming more and more assimilated to the Aryans, and their race-distinction promises completely to die out in the near future. In regard to government and civilization, they must accept the Aryan conditions, or fall under Aryan control. There is no other alternative possible.

If we look, then, over the whole world of the future, it is to behold the almost certain dominance of the Aryan type of mankind over every region except two, which alone have held and promise to hold their own. These are the regions of Arabia, and China and Japan. In these portions alone of the whole earth do we find a national energy and the existence of conditions that seem likely to repel the Aryan advance. We may briefly glance at the possible future of man in these two regions.

Since history began, Arabia has remained in an almost unchanged condition. Militant civilization has raged for thousands of years in the surrounding regions, but Arabia has lain secure behind her deserts. Kingdoms and empires have risen and fallen everywhere around this silent peninsula; yet the waves of war have broken in baffled fury upon its shores. It has poured out its hordes to conquer the civilized world, but these have brought back no civilization to its oases. It is to-day what it was three thousand years ago, — a land defying alike the sword and the habits of the civilized world. The Egyptian, the Mongol, the Turk, and the Aryan have alike retired baffled from its borders and left it to its self-satisfied sleep of barbarism. Is this to be the story of the far future as it has been of the far past? Shall civilization never penetrate the Arabian desert, and Aryan rule and Aryan commerce stand forever checked at the edge of its deadly wall of sand?

Hardly so. Modern civilization has resources which even the desert cannot withstand. A plan to conquer the desert has already been tried in the Soudan, and a similar one in Algeria. The railroad and the water-pipe may accomplish that task in which all the armies of the past signally failed. The camel, the ship of the desert, cannot compete with the iron horse, and it is among the probabilities of the future that commerce will thus penetrate to the interior of Arabia, and rouse that sleeping land to a vital activity it has never known. Civilization can scarcely fail to make its way into the Arabian oases with their enterprising populations, Aryan influence to awaken the active-minded Arabs to a realization of the wealth which lies undeveloped around them, and the oldest of known lands to join the grand movement of mankind toward the enlightenment of the future. Civilization must and will prevail over every land which barbarism now holds in its drowsy grasp, and the deserts of the world, which have so long defied its march, may yet become the slaves of the railroad and the water-pipe.

In regard to China and Japan we have before us but a question of time. The strong practical sense of their people has been abundantly demonstrated, and they need but be made clearly to perceive the advantages of Aryan methods and habits to adopt them eagerly. Japan has already realized this fact, and is introducing the conditions of Western enlightenment with a rapidity that is one of the most remarkable phenomena in the history of mankind. Such is not the case with the Chinese. Their long conservatism and their high opinion of their intellectual and industrial superiority have hindered them from fully considering the advantages possessed by the "outside barbarians." Yet such a state of affairs cannot persist. The Chinese have the same practical sense as the Japanese; and though their acceptance of the conditions of European civilization may be a slower, it will be as sure a process. Thought has never been asleep in that old land. It has simply been moving in the unchanging round of the treadmill. If it once escapes into the broader air, the stagnant conditions of Chinese civilization must give way before it, and new laws, new industries, and new ideas make their way into that realm of primitive thought.

We are here concerned with the two peoples of mankind who are least likely to fall under Aryan domination. Were they to continue dormant, they could scarcely avoid this fate. But they are not continuing dormant, and the probability is that, ere many years have passed, both China and Japan will be in a condition to defy Aryan conquest. As they become open to Aryan ideas, however, they will become more and more open to Aryan settlement, and an enlivening influence of fresh thought and fresh blood may thus penetrate to the very central citadel of Mongolian civilization. Work and thought together cannot fail to bring the antique realm of China into line with the modern and energetic nations of the Aryan West.

When this condition is realized, the commercial activity of the Aryans will undoubtedly have a rival. The Chinese are already actively commercial, and have established themselves as merchants upon many quarters of the Pacific region. Their migratory activity is also considerable. In the future we may look forward to a more vigorous contest between Chinese and Aryans in both these particulars. But it is not likely to grow very active until after the Aryans have become firmly established in every quarter of the globe. The awakening of China must be too late to give her any large share of the prize of commercial wealth and of dominion over new lands. Where the Aryan has firmly set his foot the Chinaman can never drive him out. Nor need we look upon such a probable future activity of the Chinese race as the misfortune which Chinese emigration appears to us to-day. The Chinaman of the future will undoubtedly be a higher order of being than the Chinaman of the present. He cannot but have new ideas, new hopes, new desires, and new habits. Into his dull practicality some higher degree of the imaginative and emotional must flow from connection and perhaps race-mingling with the Aryan type of man. It will undoubtedly be a slow process to lift the Chinaman from the slough of dead thought in which he has so long lain. Yet we are dealing here with the far future; and to an industrious, practical, and thinking people everything is possible.

Such are some rapid conclusions as to the possible future relations of human races and the general conditions of mankind. Doubtless they may prove in many respects erroneous, and influences which we cannot yet foresee may arise to vary and control the movements and minglings of mankind. Yet in the past, in despite of all seemingly special and voluntary influences which have affected the course of human development, the general and involuntary have held their own. The thinking and persistently enterprising race of Aryans has moved steadily forward toward dominion in both the physical and the mental empire of the world. Starting in a narrow corner of the earth, probably on the border-line of Europe and Asia, it has spread unceasingly in all directions. The contest has been a long and bitter one. At times the impulsive force of alien races has checked and turned back the Aryan march. Yet ever the Aryan force has triumphed over these obstacles, and the march has been resumed. It is still going on with undiminished energy, and it will hardly come to a halt until it has reached the termination above indicated.

The march inward has been as persistent and energetic as the march outward. The kingdom of the mind has been invaded as vigorously as the kingdom of the earth. And the conquests in this direction have been as important as those achieved over alien man and over the opposing conditions of Nature. In this direction, indeed, human progress promises to go on with undiminished energy after the earthly domain is fully occupied, and physical expansion is definitely checked. The mental empire is a boundless one. Man may lay a girdle around the earth, but the universe stretches beyond the utmost human grasp. The kingdom of knowledge has already yielded many valuable prizes to the intellectual enterprise of Aryan man, yet it is rich with countless stores of wealth, and in this domain there is room for endless endeavor. Thought need not fear any exhaustion of the world which it has set out to conquer.

If the general conditions displayed at the earliest discoverable era of the Aryan race have manifested themselves persistently till the present time, the same may be declared in a measure of the more special conditions. The development of man has taken place under the force of the inherent conditions of his physical and mental nature, and no matter how the circumstances of history might have varied, the final result could scarcely have been different from what we find it. We have endeavored to point out in preceding sections that the primitive evolution of man led inevitably to certain political relations, there named the patriarchal and the democratic. Of these the latter was the highest in grade, and directly developed, in ancient Arya, from a preceding patriarchal condition. We find this stage clearly reached nowhere else among primitive mankind, though it was closely approached in the American Indian organization, whose early condition strikingly resembled that of the Aryans.

These two conditions of barbarian organization have worked themselves out to their ultimate in a very interesting manner. All the early empires arose under patriarchal influences and became absolute despotisms. Of these China is the only one that yet persists from archaic times.

though recent kingdoms of the same type have grown up under Mongolian influence in Persia, Turkey, and Russia. All the modern Aryan kingdoms outside of Russia and Persia are more or less democratic, and possess that primitive feature of ancient Arya, the popular assembly. Popular representation — a mouthpiece of the people in the government — is the stronghold of democracy; and to this the Aryans alone, of all the races of mankind, have ever firmly held.

It is remarkable how the primitive Aryan principle of organization has retained its force through all the centuries of war and attempted despotism, and how clearly it has established itself in the most advanced modern government. Efforts numberless have been made to overthrow it. Popular representation has been prevented, despotism established, and the aid of religious autocracy brought in to hold captive the minds of men. In Russia the ancient democratic institutions have been completely overthrown, as a result of the Mongol conquest, and replaced by a patriarchal despotism. Yet these efforts have everywhere failed. Even in Russia the democratic Aryan spirit is rising in a wave that no despotism can long withstand. In Germany the recent effort to establish paternal rule is an evident failure, and must soon succumb to the peaceful rebellion of the people. In France monarchy has vanished. In England it exists only on sufferance of the representatives of the people. But in America alone can the ancient Aryan principle be said to have fully declared itself, and the government of the people by the people to have become permanently established.

America may be particularly referred to from the interesting lesson of human development it displays. It offers a remarkable testimony to the action of natural law in human progress, and the inevitable outworking of conditions in spite of every opposing effort or influence. In the government of the United States we possess the direct outcome of the government of ancient Arya, an unfoldment of the governing principle that grew up naturally among our remote ancestors, with as little variation in method as if it had arisen without a single opposing effort. It is the principle of decentralization in government as opposed to that of centralization. There are but two final types of government which could possibly arise, no matter how many intermediate experiments were made. These are the centralized and the decentralized, the patriarchal and the democratic. To the persistence of the former it is necessary that the ruler shall be at once political and religious despot. He must sway the minds of his people, or he will gradually lose his absolute control over their bodies. In China alone does this condition fully exist, and to it is due the long persistence of the Chinese form of government. In all the Aryan despotisms of to-day the autocratic rule can only persist during the continued ignorance of the people. In none of them is the emperor a spiritual potentate. With the awaking of general intelligence free government must come.

The Aryan principle of government is that of decentralization. And as no Aryan political ruler has ever succeeded in becoming the acknowledged religious head of his people, every effort at despotic centralization has failed or must fail. Local self-government was the principle of rule in ancient Arya, and it is the principle in modern America. There the family was the unit of the government. With its domestic relations no official dared interfere. The village had its governmental organization for the control of the external relations of its families, under the rule of the people. The later institution of the tribe had to do merely with the external relations of the villages; it could not meddle with their internal affairs.

As we have said, this principle has been remarkably persistent. It unfolded with hardly a check in Greece. In the Aryan village two relations of organization existed, — the family and the territorial. In Greece the former of these first declared itself, and Greek political society became divided into the family, the gens, the tribe, and the State. The family idea was the ruling principle of organization. It proved, however, in the development of civilization, to be unsuited to the needs of an advanced government, and it was replaced by the territorial idea. This gave rise to the rigidly democratic government of later Attica. It was composed of successive self-governing units, ranging downward through State, tribe, township, and family, while the people held absolute control alike of their private and their public interests. At a later date the growth of political wisdom carried this principle one step farther forward, and a league or confederacy of Grecian States was formed. Unfortunately this early outgrowth of the Aryan principle was possible in city life alone. Country life and country thought moved more slowly, and the world had to await, during two thousand years of anarchy and misgovernment, the establishment of popular government over city and country alike.

In the United States of America the Grecian commonwealth has come again to life, and the vital Aryan principle has risen to supremacy. We have here, in a great nation, almost an exact counterpart of the small

Grecian confederacy. The family still exists as the unit element, though no longer as a despotism. Then come successively the ward or the borough, the city or the township, and the county. Over these extends the State, and over all, the confederacy or United States. In each and all of these the voice of the people is the governing element. And in each, self-control of all its internal interests is, or is in steady process of becoming, the admitted principle. It is the law of decentralization carried to its ultimate, each of the successively larger units of the government having control of the interests which affect it as a whole, but having no right to meddle with interests that affect solely the population of any of the minor units.

Such is the highest condition of political organization yet reached by mankind. It is in the direct line of natural political evolution. And this evolution has certainly not reached its ultimate. It must in the future go on to the formation of yet larger units, confederacies of confederacies, until finally the whole of mankind shall become one great republic, all general affairs being controlled by a parliament of the nations, and popular self-government being everywhere the rule.

This may seem somewhat visionary. Yet Nature is not visionary, and Nature has declared, in a continuous course of events, reaching over thousands of years, that there is but one true line of political evolution. Natural law may be temporarily set aside, but it cannot be permanently abrogated. It may be hundreds, but can hardly be thousands of years before the *finale* is reached; yet however long it may take, but one end can come, — that of the confederacy of mankind. The type of government that naturally arose in the village of ancient Arya must be the final type of government of the world.

One highly important result must attend this ultimate condition, — namely, the abolition of war; for the basic principle of republican government is that of the yielding of private in favor of general interests, and the submission of all hostile questions to the arbitrament of courts and parliaments. Abundant questions rise in America which might result in war, were not this more rational method for the settlement of disputes in satisfactory operation. In several minor and in one great instance in American history an appeal has been made from the decision of the people to that of the sword. But with every such effort the principle of rule by law and by the ballot has become more firmly established, and admission of this principle is becoming more and more general as time goes on.

Unfortunately, in the world at large no such method exists for arranging the relations of states, and many wars have arisen over disputes which could satisfactorily have been settled by a congress. This is being more and more clearly recognized in Europe, and a partial and unacknowledged confederacy of the European States may be said to exist already. But the only distinct and declared avoidance of war by parliamentary action was that of the Alabama Commission, which satisfactorily settled a dispute which otherwise might have resulted in a ruinous war between America and England. This principle of confederacy and parliamentary action for the decision of international questions is young as yet, but it is growing. One final result alone can come from it, — a general confederacy of the nations, becoming continually closer, must arise, and war must die out. For the time will inevitably come when the great body of confederated nations will take the dragon of war by the throat and crush the last remains of life out of its detestable body. We can dimly see in the far future a period when war will not be permitted, when the great compound of civilized nations will sternly forbid this irrational, ruinous, and terrible method of settling national disputes, and will not look quietly on at the destruction of human life and of the results of human industry, or the wasteful diversion of industry to the manufacture of instruments of devastation. When that age comes, all hostile disputants will be forced to submit their questions to parliamentary arbitration, and to abide by the result as individuals submit to-day to the decision of courts of law. All civilized men and nations of the far future will doubtless deem it utter madness to seek to settle a dispute or reach the solution of an argument by killing one another, and will be more likely to shut up the warrior in an insane asylum than to put a sword in his hand and suffer him to run amuck like a frantic Malay swordsman through the swarming hosts of industry. Such we may with some assurance look forward to as the *finale* of Aryan political development.

Religiously the antique Aryan principle has similarly declared itself. Religious decentralization was the condition of worship in ancient Arya, and this condition has reappeared in modern America. The right of private thought and private opinion has become fully established after a hard battle with the principle of religious autocracy, and to-day every man in America is privileged to be his own priest, and to think and worship as he will, irrespective of any voice of authority.

In moral development the Aryan nations are steadily

progressing. The code of Christ is the accepted code in nearly all Aryan lands. It is not only the highest code ever promulgated, but it is impossible to conceive of a superior rule of moral conduct. At its basis lies the principle of universal human sympathy, — that of interest in and activity for the good of others, without thought of self-advantage. Nowhere else does so elevated a code of morals exist, for in every other code the hope of reward is held out as an inducement to the performance of good acts. The idea is a low one, and it has yielded low results. The idea of unselfish benevolence, and of a practical acceptance of the dogma of the universal brotherhood of mankind, is a high one, and it is yielding steadily higher results. Aryan benevolence is loftier in its grade and far less contracted in its out-reach than that of any other race of mankind; and Aryan moral belief and action reach far above those displayed by the Confucian, Buddhistic, and Mohammedan sectaries.

Industrially the Aryans have made a progress almost infinitely beyond that of other races. The development of the fruitfulness of the soil; the employment of the energies of Nature to perform the labors of man; the extensive invention of labor-saving machinery; the unfoldment of the scientific principles that underlie industrial operations, and of the laws of political economy and finance, — are doing and must continue to do much for the amelioration of man. It is not with the sword that the Aryans will yet conquer the earth, but with the plough and the tool of the artisan. The Aryan may go out to conquer and possess; but it will be with peace, plenty, and prosperity in his hand, and under his awakening touch the whole earth shall yet "bud and blossom as the rose."

There is but one more matter at which we need glance in conclusion. In original Arya the industrial organization was communistic. Yet we must look upon this as but a transitional state, a necessary stage in the evolution of human institutions. In the savage period private property had no existence beyond that of mere personal weapons, clothing, and ornaments. In the pastoral period it had little more, since the herds, which formed the wealth of the people, were held for the good of all; there was no personal property in lands, and household possessions were of small value. In the village period, though the bulk of the land was still common property, yet the house-lot, the dwelling, and its contents were family possessions. The idea of and the claim to private property has ever since been growing, and has formed one of the most important instigating elements in the development of mankind. This idea has to-day become supreme; the only general communism remaining is in government property, and the principle of individualism is dominant alike in politics, religion, and industry. Such a progressive development of individualism seems the natural process of human evolution. The most stagnant institution yet existing on the earth is the communistic Aryan village. The progress of mankind has yielded and been largely due to the establishment of the right to private property. Nor can we believe that this right will ever be abrogated, and the stream of human events turn and flow backward toward its source. The final solution of the problem of property-holding cannot yet be predicted, but it can scarcely be that of complete communism or socialism. The wheels of the world will cease to turn if ever individual enterprise becomes useless to mankind.

Yet that individualism has attained too great a dominance through the subversion of natural law by force, fraud, and the power of position, may safely be declared. Individualism has become autocratic over the kingdom of industry, and Aryan blood will always revolt against autocracy. In the world of the future some more equitable distribution of the products of industry must and will be made. The methods of this distribution no one can yet declare; but the revolt against the present inequitable condition of affairs is general and threatening. This condition is not the result of a natural evolution, but of that prevalence of war which long permitted force to triumph over right, and which has transmitted to the present time, as governing ideas of the world, many of the lessons learned during the reign of the sword. The beginning of the empire of peace seems now at hand, and the masses of mankind are everywhere rising in rebellion against these force-inaugurated ideas. When the people rise in earnest, false conditions must give way. But it is a peaceful revolution that is in progress, and the revolutions of peace are much slower, though not less sure, than those of war. The final result will in all probability be some condition intermediate between the two extremes. On the one hand, inordinate power and inordinate wealth must cease to exist and oppress the masses of mankind. On the other hand, absolute equality in station and possessions is incompatible with a high state of civilization and progress. It belongs, in the story of human development, to the savage stage of existence, and has been steadily grown away from as man has advanced in civilization. The inequalities of man in physical and mental powers are of natural origin, and must inevitably find some expression in the natural organization of society. They cannot fail to yield a certain inequality in wealth, position, and social relations. We can no more suppress this outcome of natural conditions than we can force the seeds of the oak, pine, and other forest trees alike to produce blades of grass. Enforced equality is unnatural, in that it is opposed to the natural inequalities of the body and mind of man, and it could not be maintained, though a hundred times enacted. And the inevitable tendency of even its temporary prevalence would be to check progress and endeavor, and to force human society back toward that primitive stage in which alone absolute communism is natural and possible. To find complete equality in animal relations we must go to those low forms of animal life in which there is no discoverable difference in powers and properties. The moment differences in natural powers appear, differences in condition arise; and the whole tendency of animal evolution has been toward a steadily increasing diversity of powers and faculties, until to-day there exist greater differences in this respect in the human race than at any previous period in history. These mental and physical differences cannot fail to yield social, political, and industrial diversities, though laws by the score or by the thousand should be enacted to suppress their natural influence upon human institutions.

But the existing and growing inequality in wealth and position is equally out of consonance with the lessons of Nature, since it is much in excess of that which exists in human minds and bodies, and is in numerous cases not the result of ability, but of fraud, of special advantages in the accumulation of wealth, or of an excessive development of the principle of inheritance. This evil must be cured. How, or by what medicine, it is not easy to declare. No man has a natural right to a position in society which his own powers have not enabled him to win, nor to the possession of wealth, authority, or influence which is excessively beyond that due to his native superiority of intellect. That a greater equality in the distribution of wealth than now exists will prevail in the future can scarcely be questioned, in view of the growing determination of the masses of mankind to bring to an end the present state of affairs. That the existing degree of communism will develop until the great products of human thought, industry, and art shall cease to be private property, and become free to the public in libraries, museums, and lecture-halls, is equally among the things to be desired and expected. But that superior intellect shall cease to win superior prizes in the "natural selection" of society, is a theory too averse to the teachings of Nature and the evident principles and methods of social evolution ever to come into practical realization in the history of mankind.

INDEX.

University Press: John Wilson & Son, Cambridge

CPSIA information can be obtained at www.ICGtesting.com
Printed in the USA
LVOW04s2140110813

347356LV00001B/81/A